Applied Programmable Logic Controllers Laboratory Manual

Daniel H. Nichols

THOMSON
DELMAR LEARNING

Australia Canada Mexico Singapore Spain United Kingdom United States

Applied Programmable Logic Controllers Laboratory Manual

Daniel H. Nichols

Vice President, Technology and Trades SBU:
Alar Elken

Editorial Director:
Sandy Clark

Senior Acquisitions Editor:
Stephen Helba

Development:
Dawn Daugherty

Marketing Director:
David Garza

Channel Manager:
Dennis Williams

Marketing Coordinator:
Stacey Wiktorek

Production Director:
Mary Ellen Black

Production Manager:
Larry Main

Production Coordinator:
Kara A. DiCaterino

Art/Design Coordinator:
Francis Hogan

Printed in the United States of America
1 2 3 4 5 XX 07 06 05

For more information contact
Thomson Delmar Learning
Executive Woods
5 Maxwell Drive, PO Box 8007,
Clifton Park, NY 12065-8007
Or find us on the World Wide Web at
www.delmarlearning.com

Library of Congress Cataloging-in-Publication Data:
Card Number:

ISBN: 1-4018-9967-6

NOTICE TO THE READER

Publisher does not warrant or guarantee any of the products described herein or perform any independent analysis in connection with any of the product information contained herein. Publisher does not assume, and expressly disclaims, any obligation to obtain and include information other than that provided to it by the manufacturer.

The reader is expressly warned to consider and adopt all safety precautions that might be indicated by the activities herein and to avoid all potential hazards. By following the instructions contained herein, the reader willingly assumes all risks in connection with such instructions.

The publisher makes no representation or warranties of any kind, including but not limited to, the warranties of fitness for particular purpose or merchantability, nor are any such representations implied with respect to the material set forth herein, and the publisher takes no responsibility with respect to such material. The publisher shall not be liable for any special, consequential, or exemplary damages resulting, in whole or part, from the readers' use of, or reliance upon, this material.

Contents

SECTION 7 Motor Control

SECTION 8 PLCs with Analog to Digital and Digital to Analog Converters

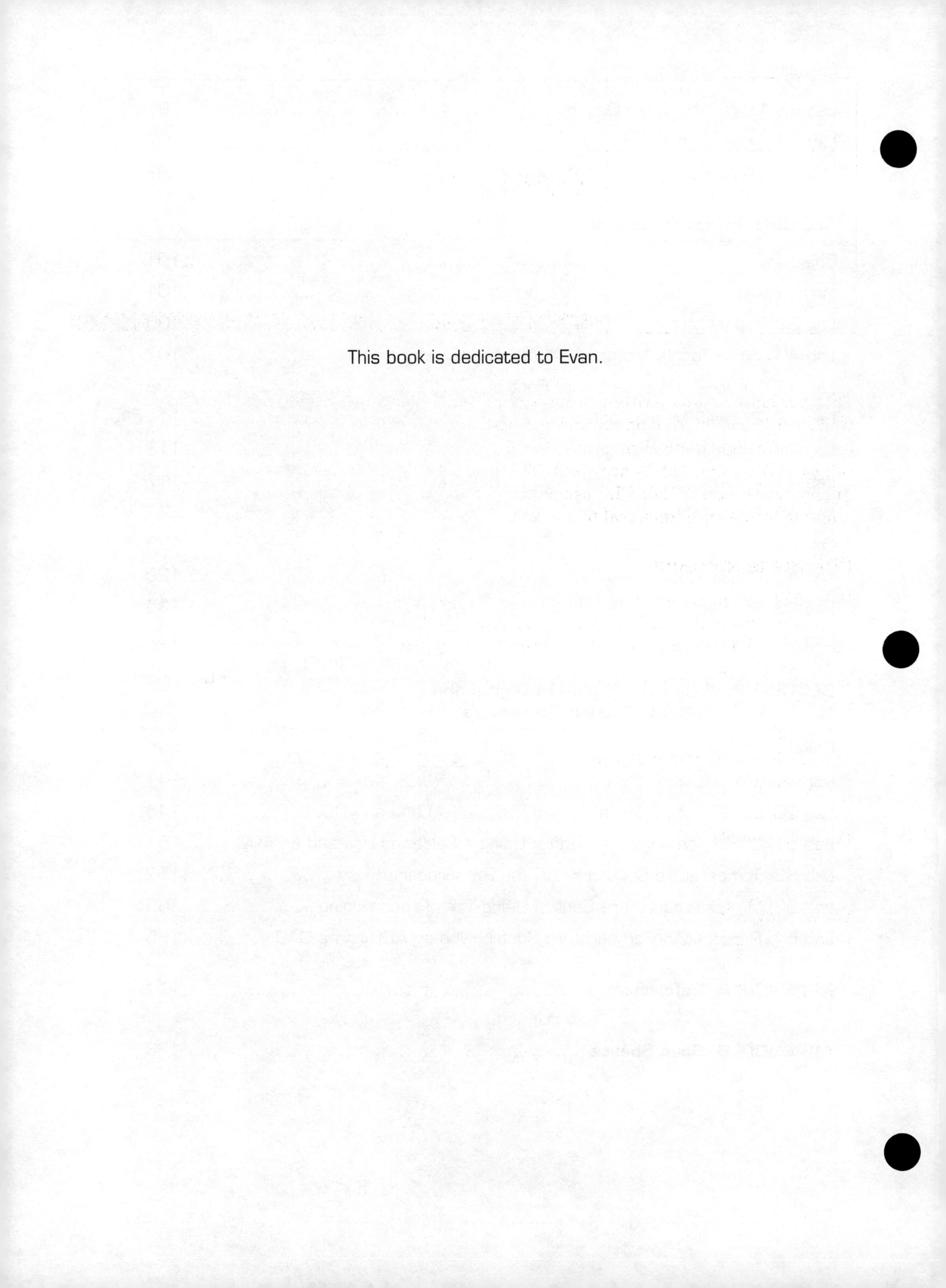

This book is dedicated to Evan.

Preface

This lab manual was written in an attempt make learning programmable logic controllers (PLCs) fun. Making connections with familiar control systems, such as conveyer belts, traffic lights, etc., brings out the how and why latches, counters, timers, sensors, relays, ADCs, and DACs are used. The programs in this manual were written for AutomationDirect's® DL05 PLC using their DirectSoft32® software, however the ideas and program layouts are universal to any PLC.

Acknowledgments

I would like to thank two of my colleagues at DeVry University Chicago, Don Ingram for his advice on some of the PLC programs, and Patrick O'Conner for his drawings in the introduction to ladder logic section of this book.

The author and Thomson Delmar Learning would like to thank the reviewers for their many suggestions and helpful comments:

James Blackett
Thomas Nelson Community College
Hampton, VA

Craig Waldvogel
DeVry University
Addison, IL

Ricardo Unglaub
DeVry University
Colorado Springs, CO

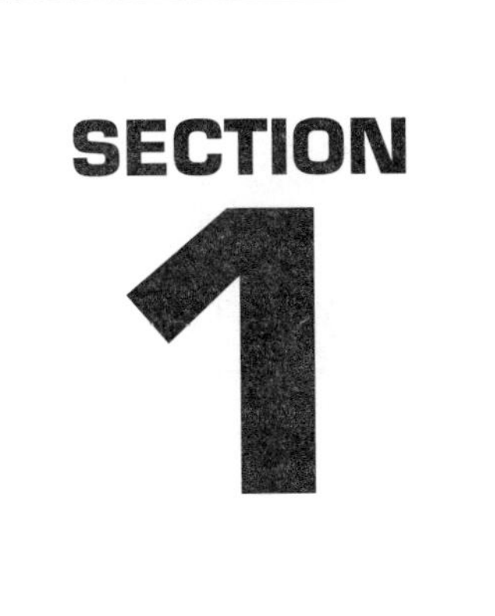

Introduction to Ladder Logic Diagrams and Programmable Logic Controllers

Ladder logic diagrams, like electronic schematics, are drawings that represent an electronic circuit. Unlike electronic schematics however, ladder logic diagrams are drawn in a vertical column format similar to a ladder. The left rail is considered the "hot" side (120V AC, 12V DC, etc.), the right rail is the neutral or ground. Switches are connected to the "hot" side and then to a device they are to control.

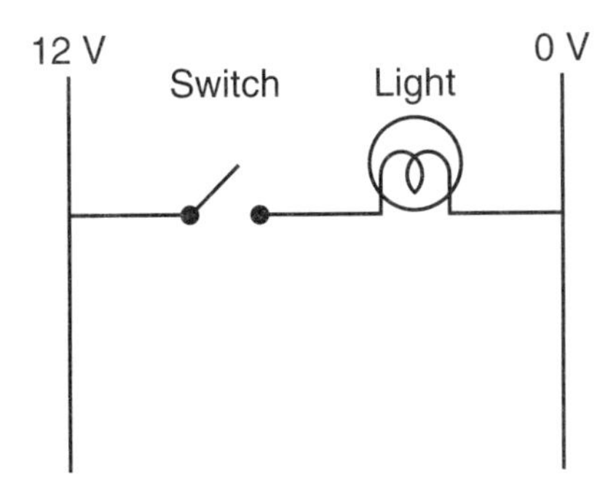

Figure 1.1 A ladder logic diagram. The left rail is considered the "hot" side (120V AC, 12V DC, etc.), the right rail is the neutral or ground.

In ladder logic, switches (like these push-buttons) are drawn like capacitors.

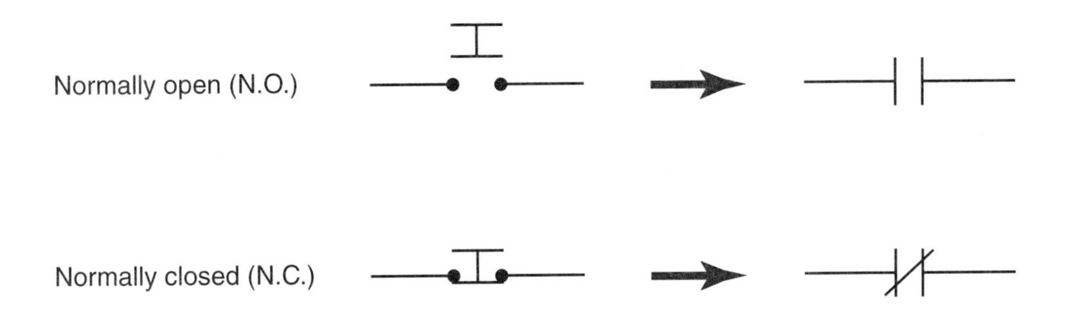

Figure 1.2 Switch contacts are drawn like capacitors.

In Figure 1.3, the equivalence of the two circuit drawing conventions are shown.

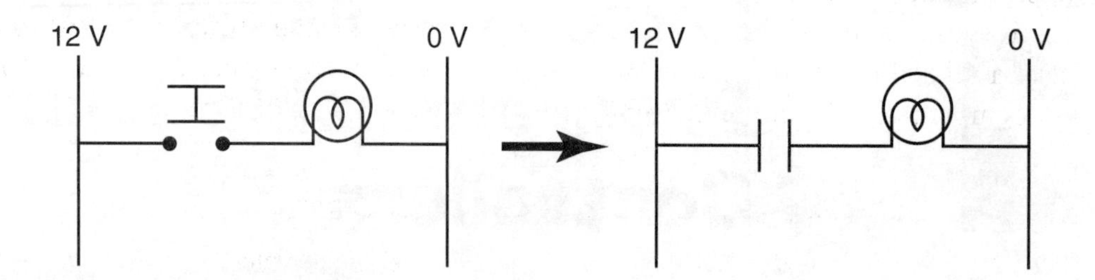

Figure 1.3 These two circuits are equivalent.

Programmable logic controllers (PLCs) use ladder logic programmed by the user to control equipment, i.e., motors, lights, etc. Every PLC has a set of inputs and outputs. The inputs are used to activate the switches in the software, which turn on the outputs. Input and output connections are made on the PLC. Not all PLCs have the same notation labeling inputs and outputs. In this manual we will represent inputs with the letter X and outputs with the letter Y. Figure 1.4 illustrates a normally open switch connected to a light.

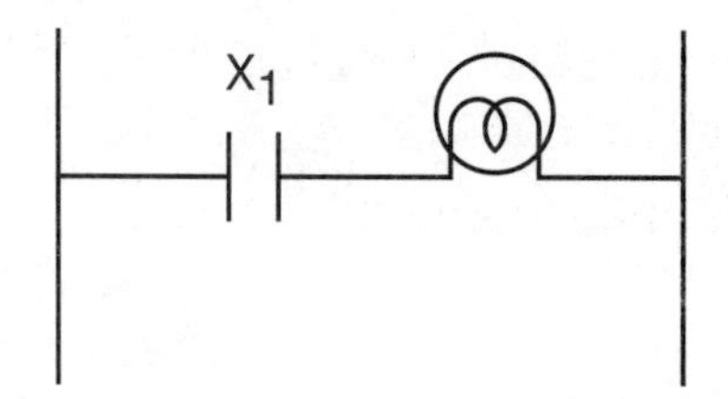

Figure 1.4 If contact X1 is closed, the light will turn on.

A PLC can be programmed so that when certain inputs are activated programmed outputs will be turned on. To illustrate how the simple circuit shown in Figure 1.4 can be realized using a PLC, look at the Figure 1.5. In Figure 1.5, SW1 is connected to input X1. A light is connected to output Y1. When mechanical SW1 is closed, 12V is applied to X1. This tells the software to close switch X1 in the program, which turns on output Y1 and hence the light connected to output Y1.

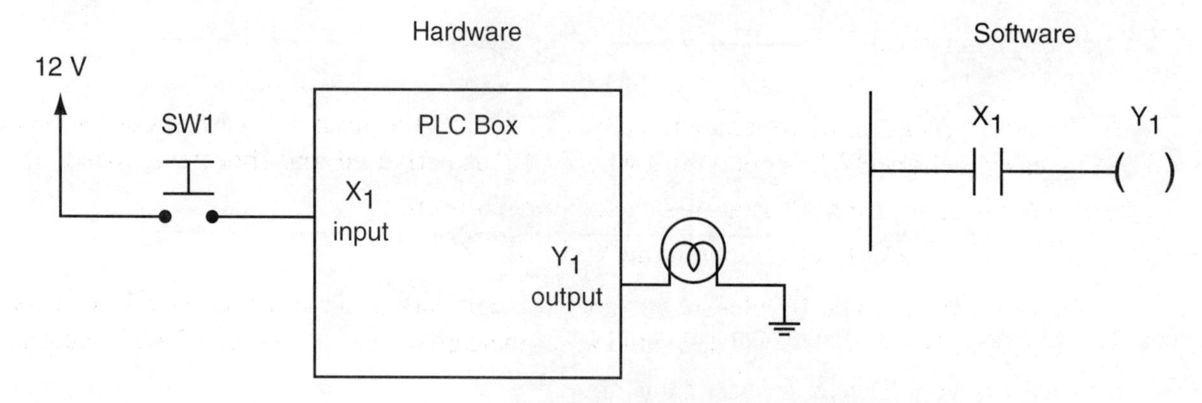

Figure 1.5 The PLC software determines what output to turn on when a given input is activated. When SW1 is closed, contact X1 in the program will close and Y1 will turn on.

A normally open switch is abbreviated N.O. and normally closed switch N.C. To activate a switch means to change it to its opposite state. A normally open switch is open until someone activates it (closes it). A normally closed switch is closed until someone activates it (opens it). In all schematic diagrams switches are shown in their inactive state, either normally open or normally closed. See Figure 1.6.

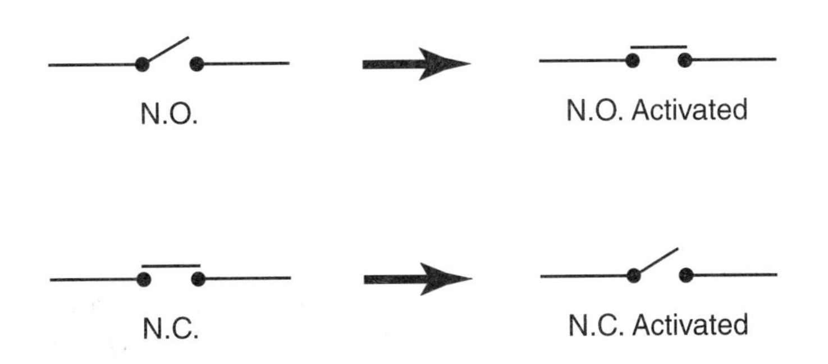

Figure 1.6 To activate a switch means to change it to its opposite state, from normally open (N.O.) to closed or from normally closed (N.C.) to open.

In Figure 1.7, a normally open switch SW1 controls a normally open X1. Let us look at the two cases: where SW1 is open, and where SW1 is activated and therefore closed.

a. If SW1 is open, then X1 remains open and Y0 is off.

b. If SW1 is closed, then X1 closes and Y0 is on.

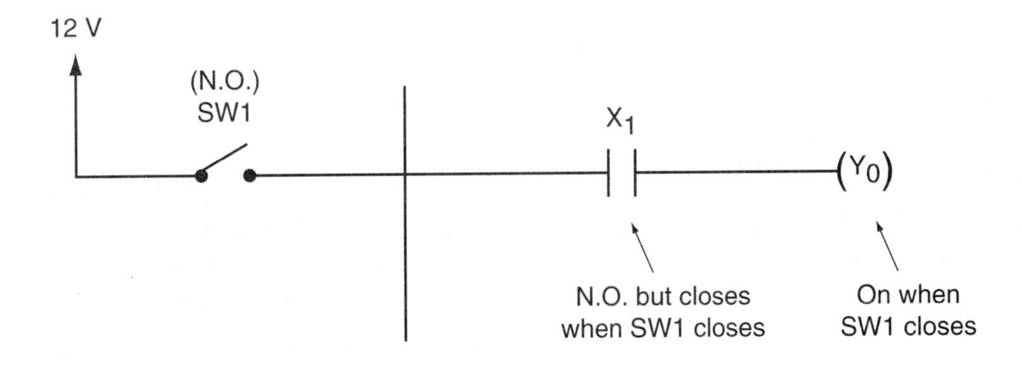

Figure 1.7 If SW1 is open, then X1 remains open and Y0 is off. If SW1 is closed, then X1 closes and Y0 is on.

In Figure 1.8, a normally open switch SW1 controls a normally closed X1. Let us look at the two cases: where SW1 is open, and where SW1 is activated and therefore closed.

a. If SW1 is open, then X1 remains closed and Y0 is on.

b. If SW1 is closed, then X1 opens and Y0 is off.

In Figure 1.9, a normally closed switch SW1 controls a normally open X1. Let us look at the two cases: where SW1 is activated and therefore opens, and where SW1 is closed.

a. If SW1 opens, X1 is open and Y0 is off.

b. If SW1 is closed, then X1 is closed and Y0 is on.

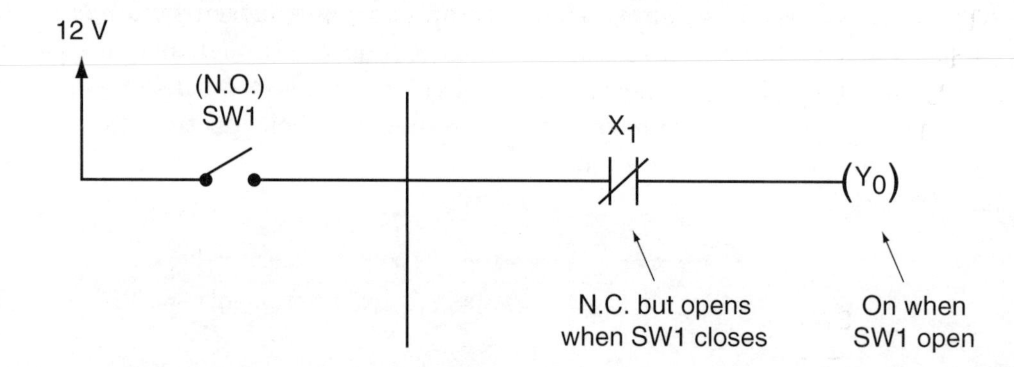

Figure 1.8 If SW1 is open, then X1 remains closed and YO is on. If SW1 is closed, then X1 opens and YO is off.

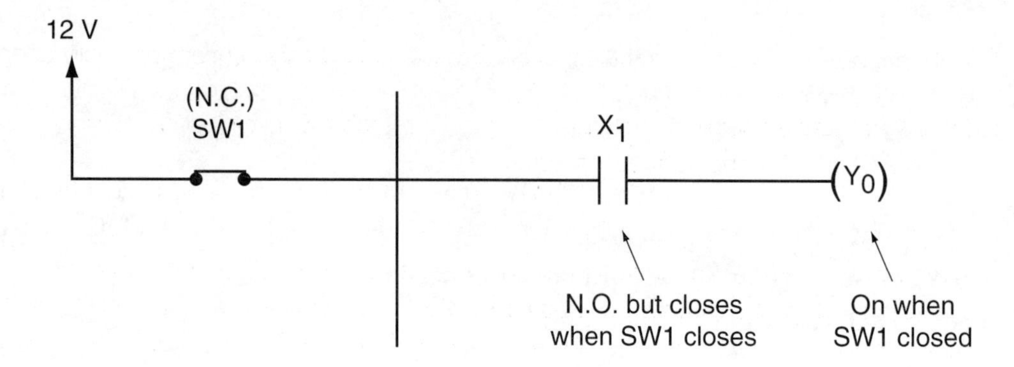

Figure 1.9 If SW1 is activated, SW1 opens, X1 is open, and YO is off. If SW1 is closed, then X1 is closed and YO is on.

In Figure 1.10 a normally closed switch SW1 controls a normally closed X1. Let us look at the two cases: where SW1 is activated and therefore opens, and where SW1 is closed.

a. If SW1 is activated, SW1 opens, X1 is closed, and Y0 is on.

b. If SW1 is closed, then X1 is open, and Y0 is off.

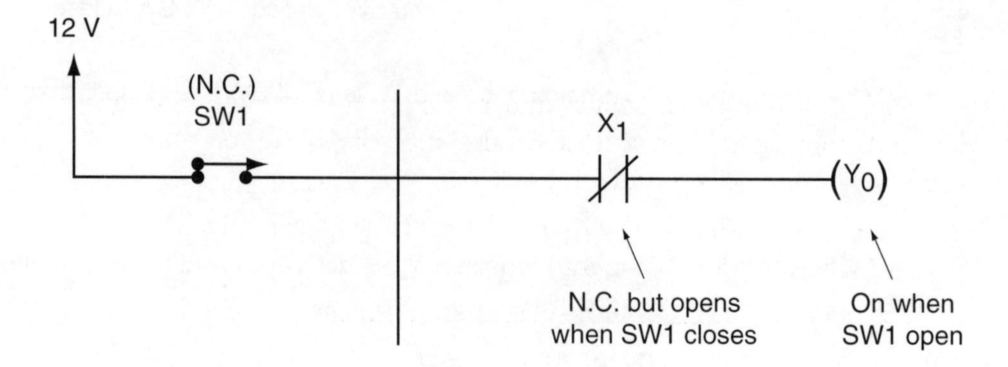

Figure 1.10 If SW1 is activated, SW1 opens, X1 is closed, and YO is on. If SW1 is closed, then X1 is open, and YO is off.

The circuits in Figures 1.7 through 1.10 have been simplified for clarity. In reality, it is a little more complicated. As the switches in the software open and close, they control the outputs. The outputs are switches themselves with a common terminal. When an output is commanded to turn on, the output switch closes, connecting it to the common. The output on the PLC can be made to output any voltage by changing the voltage on the common. See Figure 1.11.

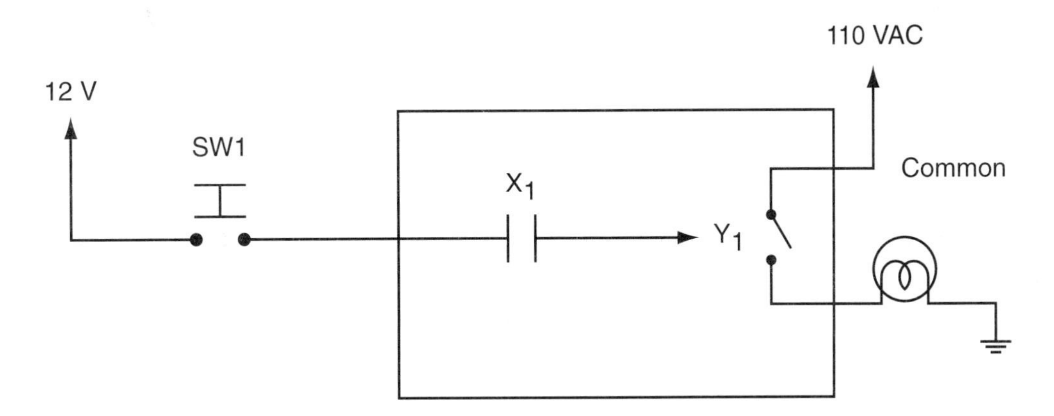

Figure 1.11 More detailed view of the PLC. If external switch SW1 is closed, internal switch X1 closes, Y1 is activated, and the light turns on.

Generally, not all PLC manufacturers use the same notation and format when labeling inputs, outputs, and the diagrams. For example, let us write the same ladder logic diagram for the simple circuit, where a single switch is controlling an output for several different manufacturers of PLCs, Figure 1.12.

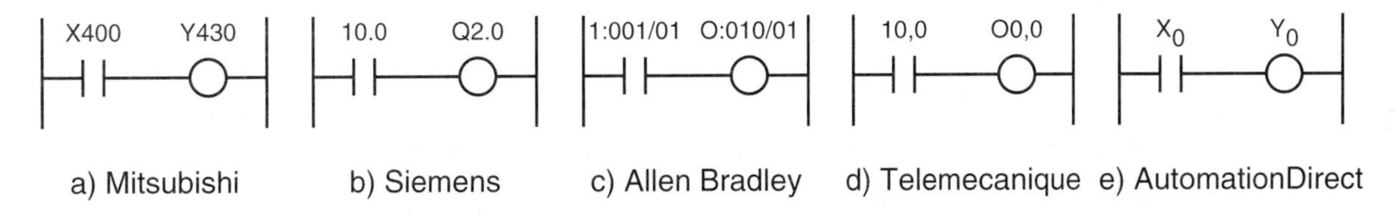

Figure 1.12 Various notation schemes for a simple circuit where a single input is controlling an output for four different manufacturers: a) Mitsubishi, b) Siemens, c) Allen Bradley, d) Telemecanique, e) AutomationDirect.

In this book, Xs are used for inputs and Ys for outputs. Labeling used in this book is consistent with AutomationDirect's® PLCs. Like Mitsubishi, AutomationDirect uses Xs for inputs and Ys for outputs. See Figure 1.13.

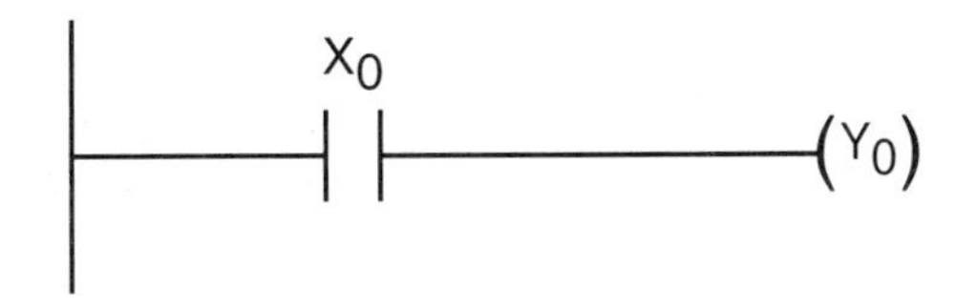

Figure 1.13 Labeling used in this book is consistent with AutomationDirect's® PLCs: Xs for inputs and Ys for outputs.

A comparison of different PLCs is give in Figure 1.14

Competitive Comparison				
	AutomationDirect DirectLOGIC 05	Allen-Bradley MicroLogix™ 1000*	GE Fanuc VersaMax Micro	Siemens S7-222
Removable terminal blocks	YES	NO	YES	YES
Number of input points	8	6	8	8
Number of output points	6	4	6	6
Communication ports	2	1	1	1
Total program/data memory	6K	1K	6K	6K
Analog I/O Option Module	YES	NO	NO	YES
Master/Slave Protocols	YES	SLAVE ONLY	SLAVE ONLY	YES
Real Time Clock and Memory Cartridge Options	YES	NO	NO	YES
PID	YES	NO	YES	YES
DeviceNET Slave Option Card	YES	YES	NO	NO
Auxiliary DC Power Supply	NO	YES	YES	YES

Allen-Bradley Micrologix 1000 Installation Instructions Manual 176151. Allen-Bradley Micrologix 1000 User Manual 176163. GE Fanuc Series VersaMax Nano & Micro Controller Solutions GFA-196 50M 2/00. Siemens Simatic S7-200 Programmable Controller System Manual IC79000-G7076-C233-02 Edition: 2 All product names, trademarks, and registered trademarks are the property of their respective manufacturers. AutomationDirect disclaims any proprietary interest in the marks and names of others.

Figure 1.14 A comparison of similar PLCs from four different manufacturers. (Courtesy of AutomationDirect.com.)

NOTE

Because of the wide variety of PLCs, specific wiring and labels (inputs, outputs, and commons) vary among them. Be sure to consult your spec sheet for your particular PLC.

Name ______________________________ Date ______________

Lab

AutomationDirect's® DL05 and DirectSOFT32®

Objective

Upon completion of this lab, you should be able to:

- Write a simple program using DirectSOFT32® and demonstrate its use with AutomationDirect's® DL05 PLC. The procedures learned for programming the DLO5 PLC in this lab will be used throughout the rest of the manual.

Introduction

AutomationDirect's® DL05 PLC has 8 inputs and 6 outputs. The common for input X0–X3 is C0, and for X4–X7 is C1. These commons should be grounded. The common for outputs Y0–Y2 is C2, and for Y3–Y5 is C3. In order to activate an input, 12VDC–24VDC has to be applied to the input. See Figure 1.15 for a spec sheet on the DL05.

Equipment Required

1. AutomationDirect's® DL05 PLC and DirectSOFT32® software
2. 12VDC power supply
3. 1kohm resistor
4. 1 LED
5. Switch

The software used to control the DL05 is called DirectSOFT32®. The following is a brief description on how to get started using it by demonstrating how to enter a simple program that turns on an LED with a switch. See Figure 1.16

Procedure

STEP 1. Begin the lab by wiring the PLC according to the wiring diagram in Figure 1.16.

STEP 2. The program will first be written on a computer using the software DirectSOFT32®. Then the program will be downloaded into the PLC and commanded to run. To begin using the DirectSOFT32® software, perform the following:

1. Click on **DSLaunch**.

DL05 I/O Specifications

D0-05DR $99.00

Wiring diagram and specifications

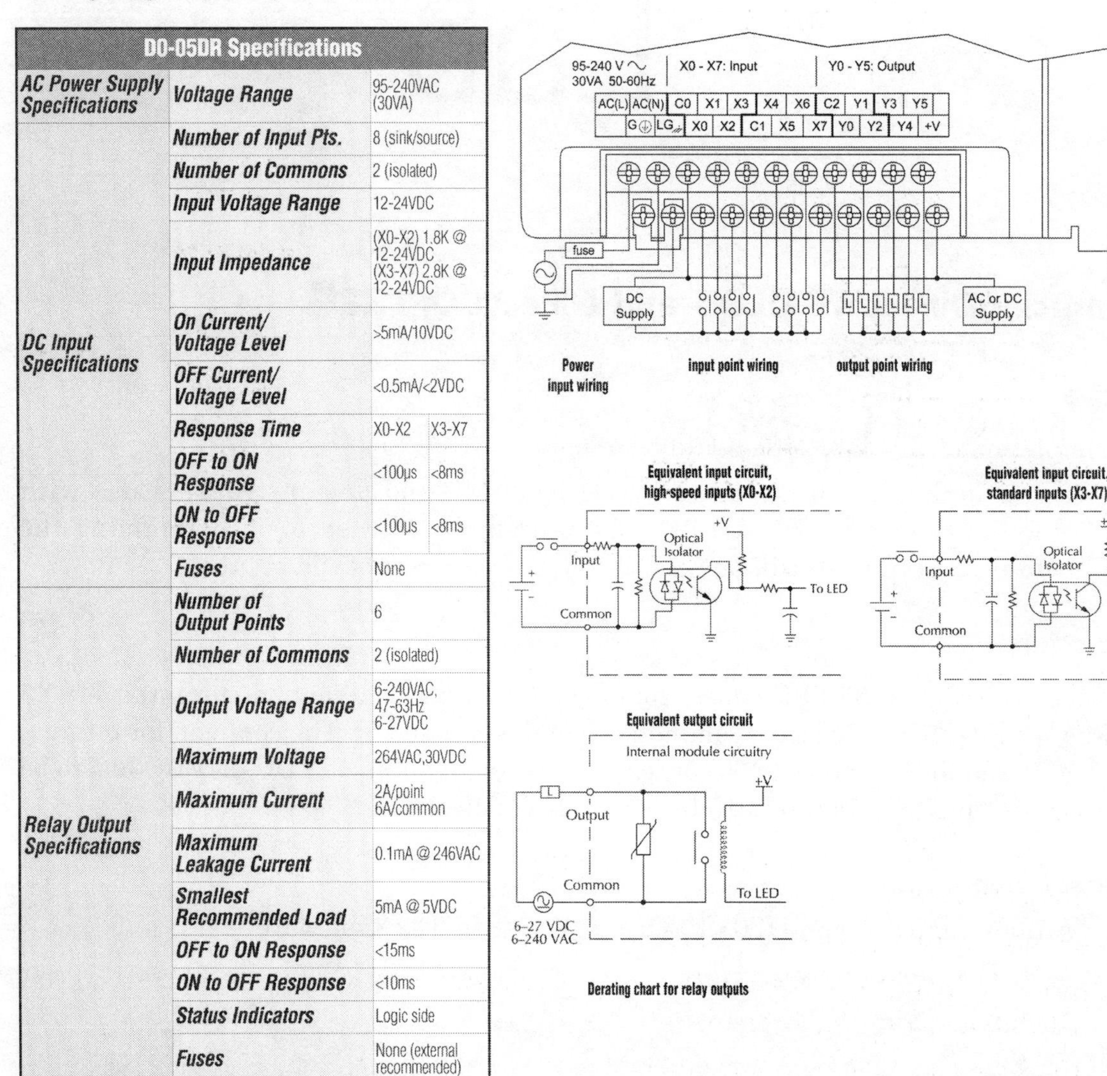

D0-05DR Specifications			
AC Power Supply Specifications	Voltage Range	95-240VAC (30VA)	
DC Input Specifications	Number of Input Pts.	8 (sink/source)	
	Number of Commons	2 (isolated)	
	Input Voltage Range	12-24VDC	
	Input Impedance	(X0-X2) 1.8K @ 12-24VDC (X3-X7) 2.8K @ 12-24VDC	
	On Current/ Voltage Level	>5mA/10VDC	
	OFF Current/ Voltage Level	<0.5mA/<2VDC	
	Response Time	X0-X2	X3-X7
	OFF to ON Response	<100µs	<8ms
	ON to OFF Response	<100µs	<8ms
	Fuses	None	
Relay Output Specifications	Number of Output Points	6	
	Number of Commons	2 (isolated)	
	Output Voltage Range	6-240VAC, 47-63Hz 6-27VDC	
	Maximum Voltage	264VAC,30VDC	
	Maximum Current	2A/point 6A/common	
	Maximum Leakage Current	0.1mA @ 246VAC	
	Smallest Recommended Load	5mA @ 5VDC	
	OFF to ON Response	<15ms	
	ON to OFF Response	<10ms	
	Status Indicators	Logic side	
	Fuses	None (external recommended)	

Typical Relay Life (Operations) at Room Temperature		
Voltage and Type of Load	Load Current 1A	2A
24 VDC Resistive	600K	270K
24 VDC Solenoid	150K	60K
110 VAC Resistive	900K	350K
110 VAC Solenoid	350K	150K
220 VAC Resistive	600K	250K
220 VAC Solenoid	200K	100K

Figure 1.15 Specifications on AutomationDirect's® DL05 PLC. (Courtesy of AutomationDirect.com.)

2. Click on **DirectSOFT32® Programming**.
3. Choose **DL05**.

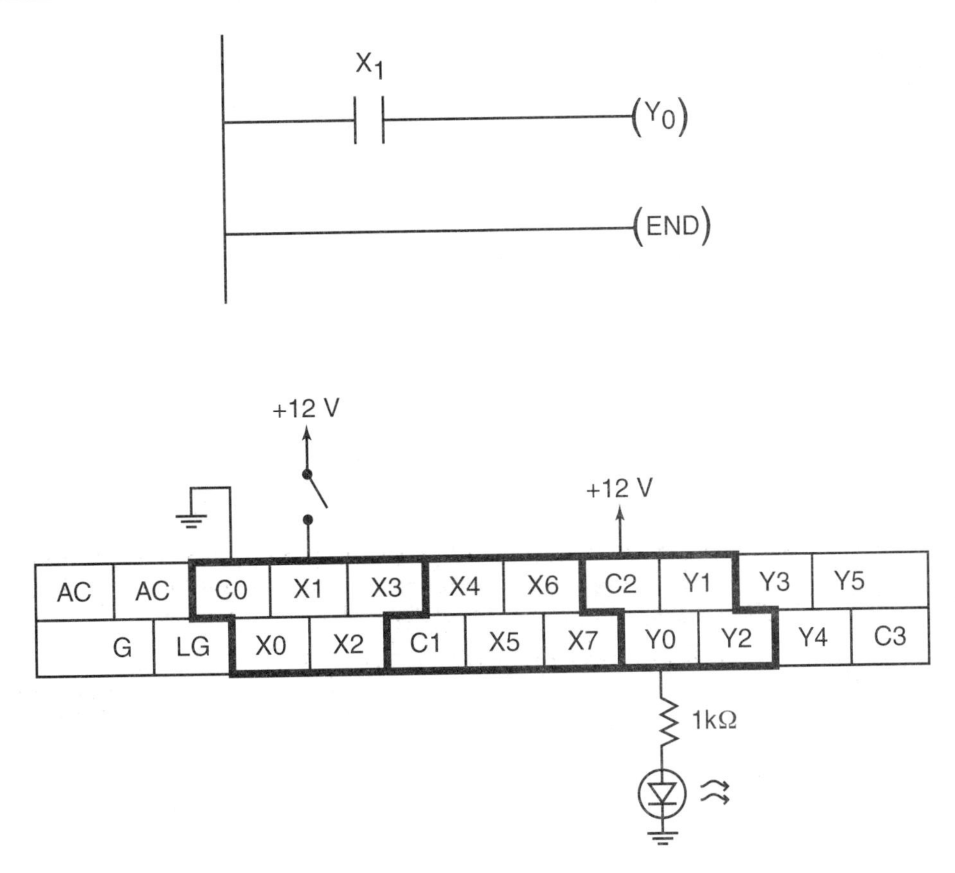

Figure 1.16 Program and wiring diagram for a simple circuit to turn on a LED connected to YO by activating X1.

4. Click on **Edit** then **Edit mode.**
5. In the program area, type in your program. The selection of switches is listed at the bottom of the screen.
6. On the last rung, type the **END** statement.
7. Click on **Edit**, then **Accept**. If there were no errors, the yellow line visible on the left hand side of the screen will turn green.
8. Click on **File**, then **Save** the program to the hard drive (a floppy will take "forever").

Downloading the Program into the PLC

Download the program into the PLC using the following steps. Refer to Figure 1.17 for connecting the PLC to power and your computer.

1. Plug in the PLC to 110VAC.
2. Plug in the cable to the back of the PLC into port 1. Plug in the other end of this cable to the COM port on the back of your computer.
3. Flip the switch on the back of the PLC to TERM, which stands for terminal.
4. Click on **PLC** on the tool bar, then click on **Connect**. Select COM1 if present; if not, click on **Add**, and follow the instructions.
5. Click on **Use Disk**.
6. Click on **File**, **Write program** to **PLC**.

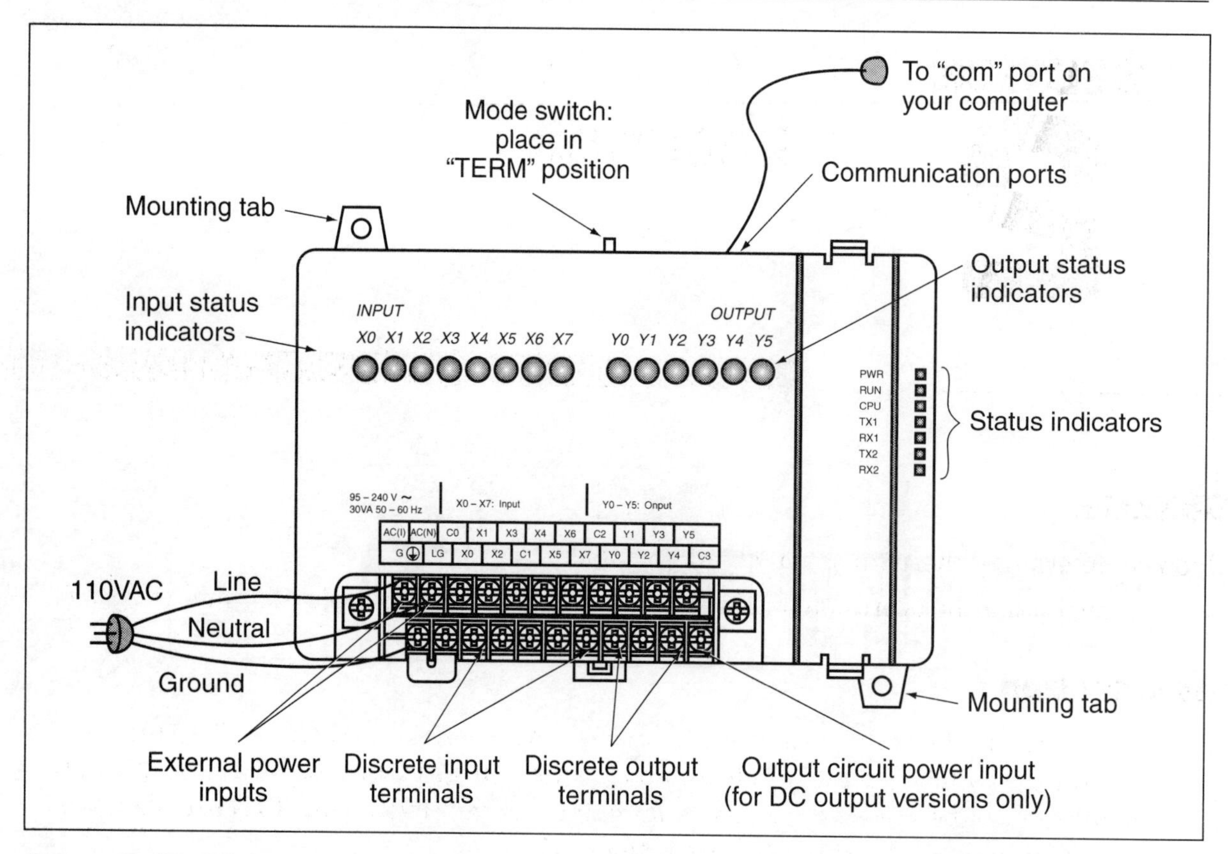

Figure 1.17 Diagram illustrating power and computer connections to the DL05 PLC.

7. Click on the "stop light" and choose **Run**.

To demonstrate the lab, apply 12VDC to X1 while observing the output.

Did Y0 and the LED turn on?

Congratulations! You have written your first PLC program.

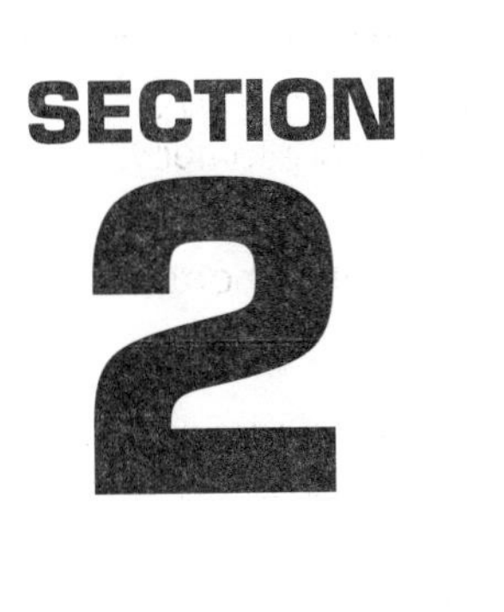

Latches

Objective

Upon completion of these labs, you should be able to:

- Understand the operation of a latch and be able to write a simple program using one.

Introduction

A latch is a circuit where the output stays on once momentarily activated. They are used with momentary switches, for example, in an elevator, a floor is selected by pressing a button. The button can be released because a latch latches that button on. To reset the latch, a reset must be activated. A latch is illustrated in Figure 2.1

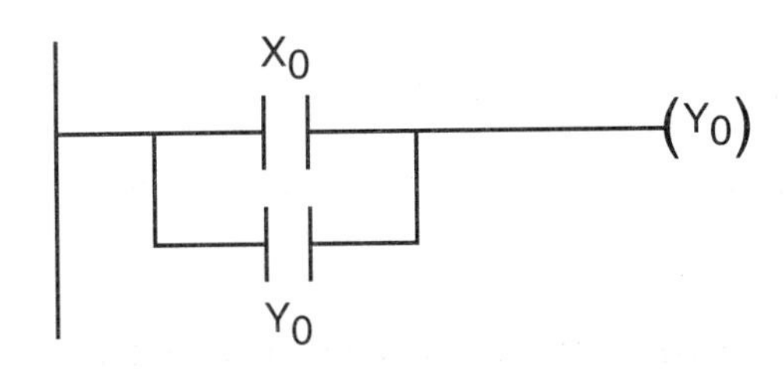

Figure 2.1 A latch.

Notice how in Figure 2.1, the output Y0 is placed across the input X0. When X0 closes, Y0 is activated, closing the switch Y0 placed across the input X0. If X0 is allowed to open, the output is still commanded to stay on because switch Y0 is closed. Notice there is no way to turn off the output once it is turned on.

To turn off a latched output a reset is used. A normally closed switch is placed in series with the latch. In Figure 2.2, X1 is used as a reset. If the circuit is latched on and X1 is activated, it will turn off the output and switch Y0 will open. This will "break" the latch and Y0 will remain off.

The Positive and Negative Differential Contacts: (LAB 4)

The positive and negative differential contacts are designed to close for one CPU scan, about 1ms when activated, as illustrated in Figure 2.3 and Figure 2.4. The positive differential will close briefly when the contact receives a low to high transition, 0V to 12V. Any output

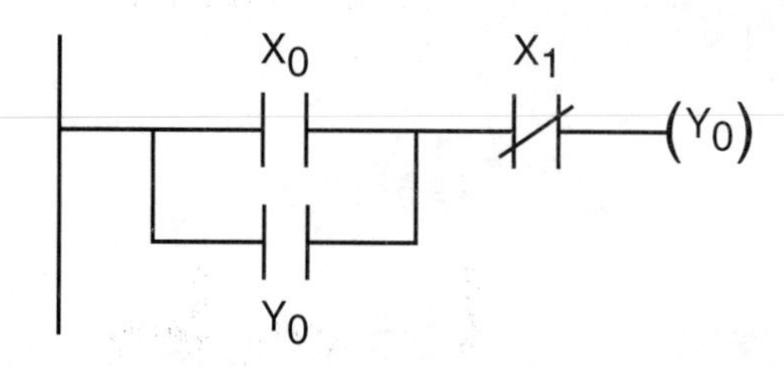

Figure 2.2 Latch with reset.

connected to it will turn on for one CPU scan. This contact is sometimes called a "one-shot" because it closes for a brief period of time, and returns to its original state. See Figure 2.3.

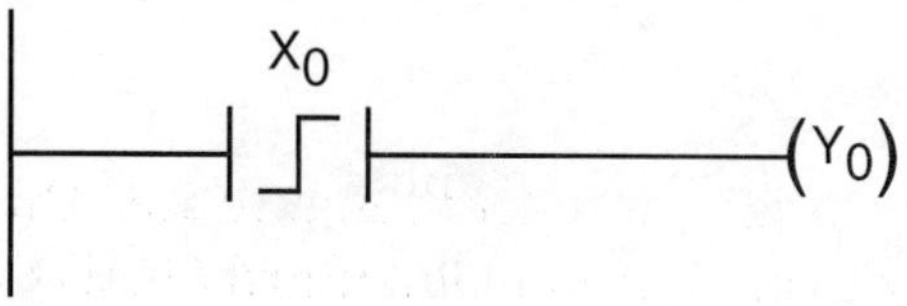

Figure 2.3 The positive differential X0 will close briefly when the contact receives a low to high transition, 0V to 12V. Y0 will turn on for one CPU scan.

The negative differential contact will close briefly when a contact receives a high to low transition, 12V to 0V. Any output connected to it will turn on for one CPU scan. See Figure 2.4.

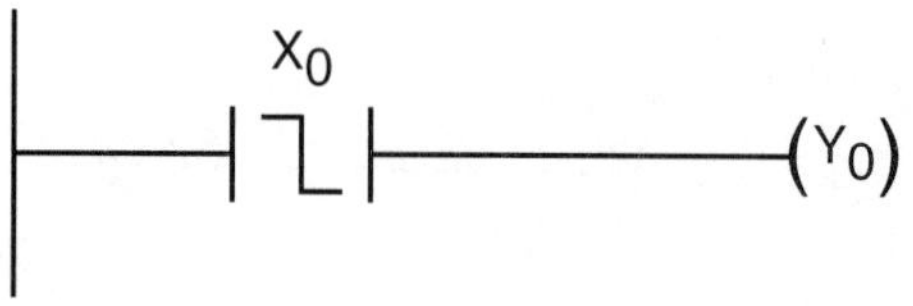

Figure 2.4 The negative differential contact X0 will close briefly when a contact receives a high to low transition, 12V to 0V. Y0 will turn on for one CPU scan.

Name ______________________ Date ______________

Lab 

Light Switches

Latches are illustrated in this program. Refer to Figure 2.5. On rung one, if X0 is activated, output Y0 turns on. On rung two, if X1 is momentarily activated, output Y1 turns on and latches. On rung three, if X3 is activated, output Y3 is turned on and latches until the reset switch X2 is activated, breaking the latch and turning off Y2.

To recap:

X0—Turns on Y0

X1—Turns on Y1

X2—Resets output Y2

X3—Turns on output Y2

Equipment Required

1. PLC
2. 12V power supply
3. LEDs
4. 1kohm resistors
5. Push button switches

To demonstrate the lab, alternately apply 12VDC to X0, X1, X2, and X3 while observing the outputs.

Questions

1. What happens to Y0 when X0 is activated momentarily?
2. What happens to Y1 when X1 is activated momentarily?
3. What happens to Y1 when X1 is activated and stays activated?
4. What is the purpose of X2 in the program?
5. What makes Y1 stay on once activated?
6. Once Y1 is turned on, how is it turned off?
7. Once Y2 is turned on, how is it turned off?

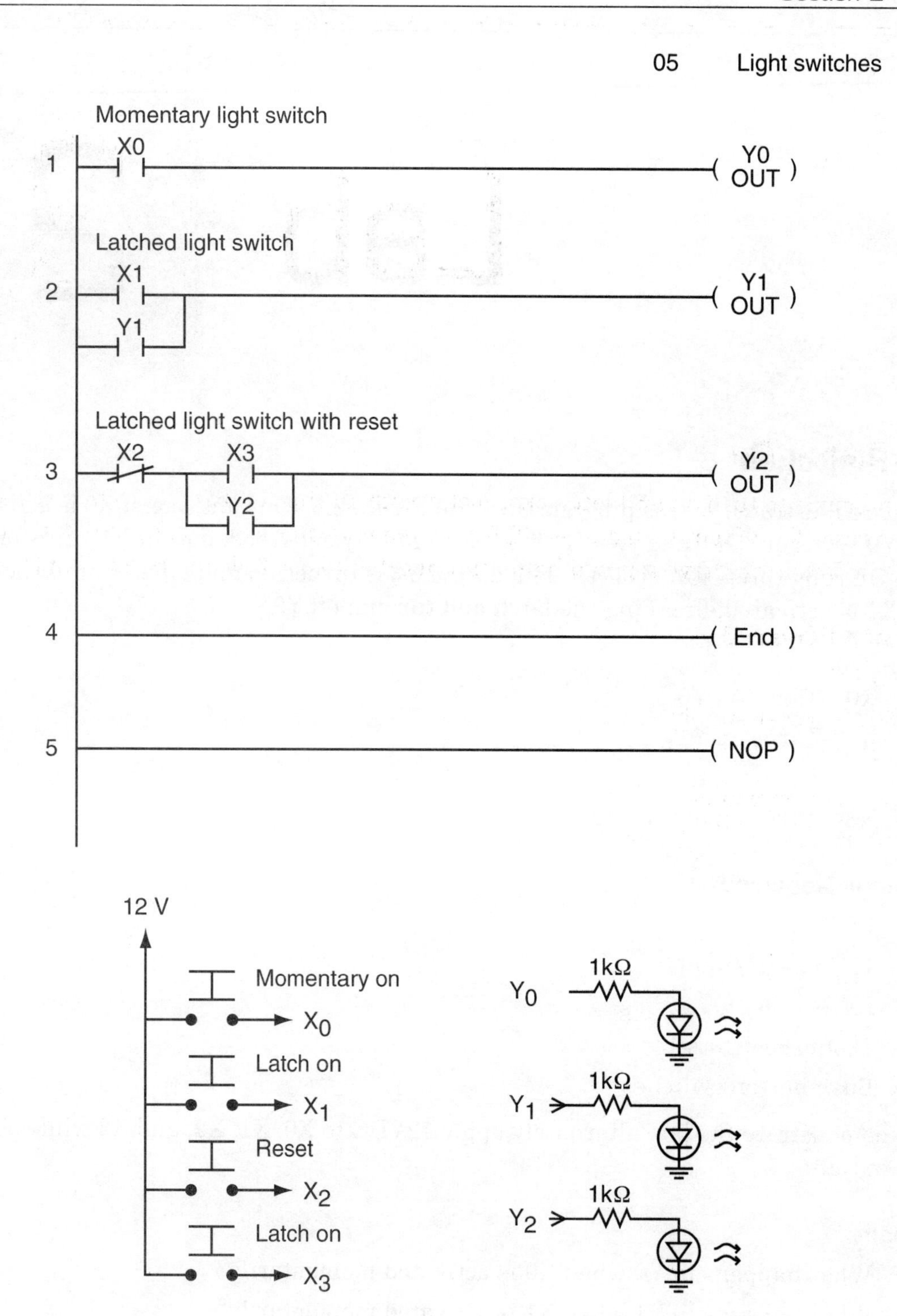
05 Light switches
Momentary light switch
1
X0
Y0
OUT
Latched light switch
2
X1
Y1
Y1
OUT
Latched light switch with reset
3
X2
X3
Y2
Y2
OUT
4
End
5
NOP
12 V
Momentary on
X0
Latch on
X1
Reset
X2
Latch on
X3
Y0
1kΩ
Y1
1kΩ
Y2
1kΩ

Figure 2.5

Name ______________________________ Date ______________

Lab

Write Latch On

Write a program to turn on and latch on a motor with an input X0 and output to the motor Y0, and a reset X1. When the motor is off, have a red indicator light on connected to Y1, and when the motor is on, have a green light on connected to Y2. Refer to Figure 2.5 for help.

Equipment Required

1. PLC
2. 12V power supply
3. LEDs
4. 1kohm resistors
5. Push button switches
6. 12VDC motor

Name ______________________________ Date ______________

Lab

Light Activated Light

An automated night light turns on a light when a room is dark. Refer to Figure 2.6. A light sensor activates X0 with a high to a low transition, corresponding to a light to dark transition. When X0 is activated, Y0 turns on and latches. When the light sensor measures a transition from dark to light, output Y1 activates for 1ms, breaking the latch on Y0, and Y0 turns off.

To recap:

X0—Turns on light with a high to low transition and turns off the light with a low to high transition

Y0—Output

Y1—1ms output breaking latch

Equipment Required

1. PLC
2. 12V power supply
3. LED
4. 1kohm resistor
5. Photo resistor
6. 10kohm pot

To demonstrate the lab, adjust the 10kohm pot while alternately covering up and uncovering the light sensor until output changes with light level.

Questions

1. What happens to Y0 when X0 makes a high to low transition?
2. What happens to Y0 when X0 makes a low to a high transition?
3. What happens to Y1 when X0 makes a low to a high transition?
4. What is the purpose of Y1 in the program?
5. What makes Y0 stay on once activated?
6. What could be wrong if the light never comes on?
7. What could be wrong if the light never turns off?

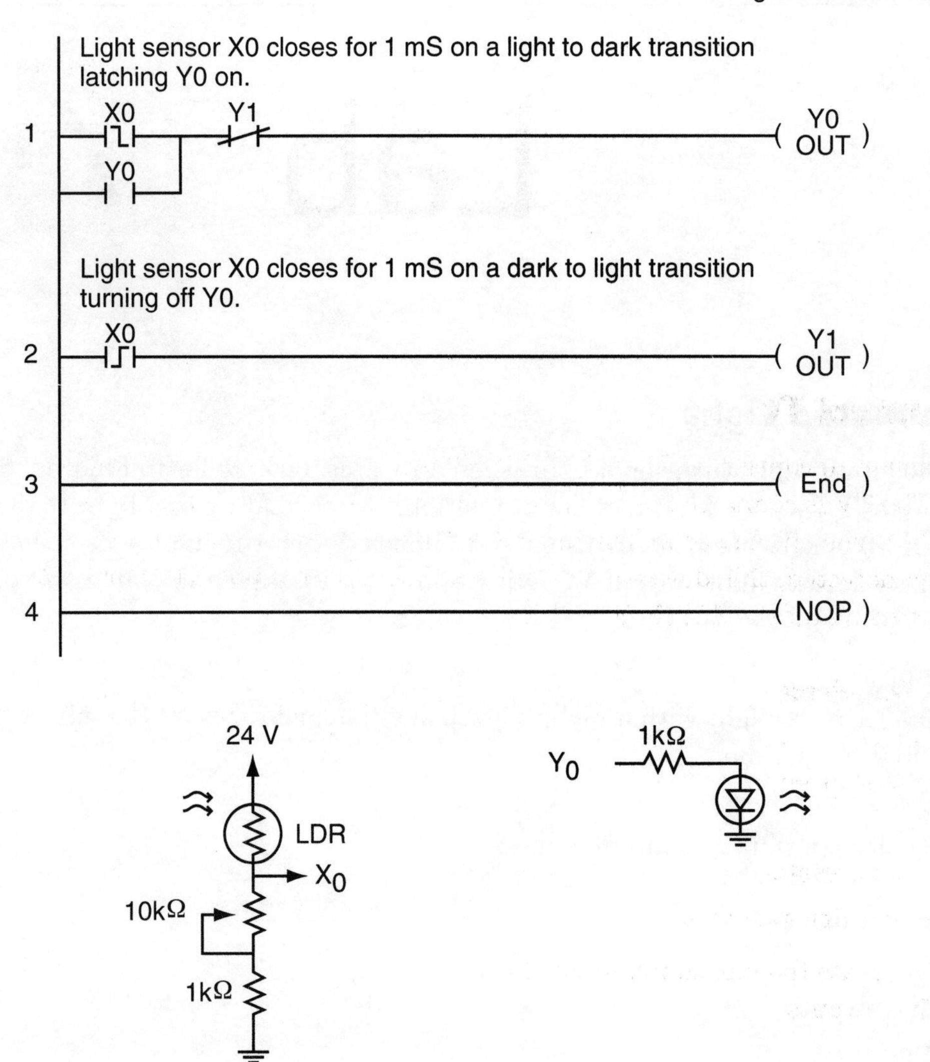
05 Light-activated Light
Light sensor X0 closes for 1 mS on a light to dark transition latching Y0 on.
1
X0
Y1
Y0
Y0
OUT
Light sensor X0 closes for 1 mS on a dark to light transition turning off Y0.
2
X0
Y1
OUT
3
End
4
NOP
24 V
LDR
X0
10kΩ
1kΩ
Y0
1kΩ

Figure 2.6

Name ______________________________ Date ______________

Lab

Write Control TV

Write a program to control a TV with a PLC. The TV turns on when X0 is activated and latches on. The TV is connected to Y0. The program is to be written so that when the phone rings, the TV turns off. A special sensing circuit listens for the frequency of a ringing telephone. When detected, this device sends out a 12V signal to input X1 to break the latch of output Y0, turning off the TV. Refer to Figure 2.6 for help.

Equipment Required

1. PLC
2. 12V power supply
3. LED
4. 1kohm resistor
5. Momentary switches

To demonstrate the lab, simulate turning the TV on and off with inputs X0 and X1 while observing the outputs.

SECTION 3

Logic Gates

Objective

Upon completion of these labs, you should be able to:

- Understand how to implement the basic logic gates AND, OR, NAND, NOR, XOR, XNOR, NOT, for control using ladder logic.

Introduction

All logic gates can be built using a combination of normally open (N.O.) switches and normally closed (N.C.) switches in series and parallel combinations.

If A **and** B are both closed, the light will go on. This is an AND gate, Figure 3.1.

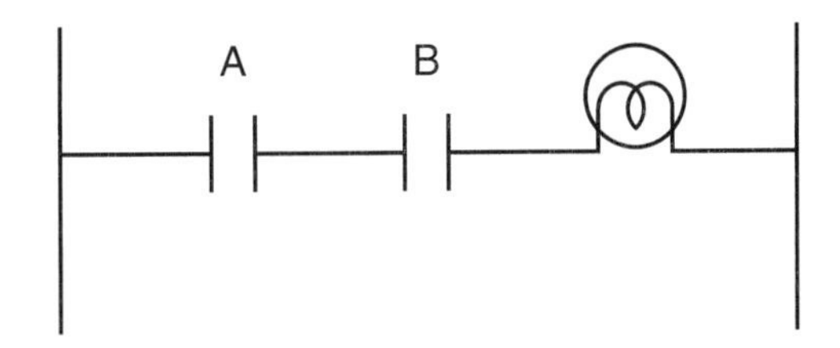

Figure 3.1 An AND Gate.

If either A **or** B are closed, the light will go on. This is an OR gate, Figure 3.2.

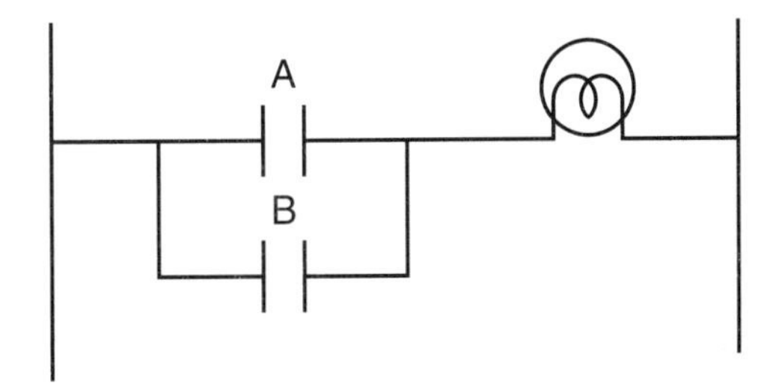

Figure 3.2 An OR Gate.

A logic true in digital electronics is 5V while for the PLC it will be 12V.

To simulate a logic true on an input, apply 12V to the input. Figure 3.3 illustrates the seven basic logic gates.

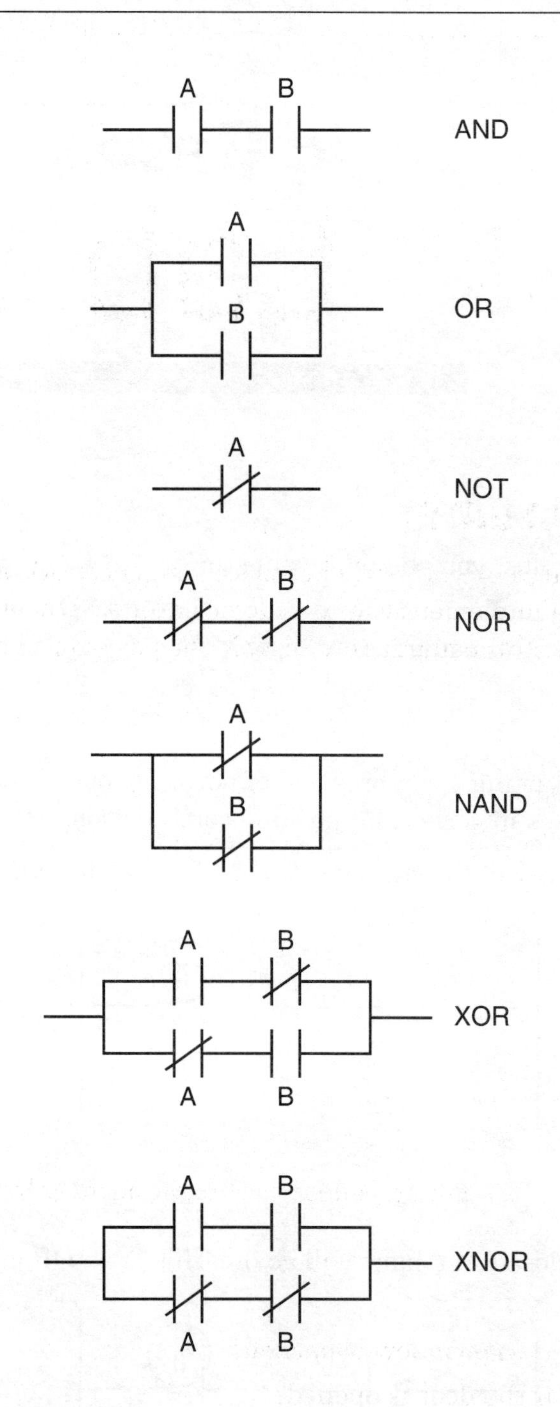

Figure 3.3 Seven basic logic gates.

Name ______________________________ Date ______________

Lab 

Burglar Alarm (OR-GATE)

A burglar alarm is activated by N.O. switches placed on a door, X1, and a window, X2. Refer to Figure 3.4. When the alarm is armed and a door or window is opened, the alarm will latch on. The only way to turn off the alarm is to disable the power or to hit the reset button, X0.

To recap:

X0—Reset

X1—Door switch

X2—Window switch

Y0—Alarm power

Equipment Required

1. PLC
2. 12V power supply
3. Push button switches
4. Buzzer

To demonstrate the lab, simulate a door or window opening by applying 12VDC to X1, X2, while observing the output.

Questions

1. What happens if the window is opened?
2. What happens if the door is opened?
3. What happens if both the door and window are opened at the same time?
4. How can the alarm be turned off?
5. What could cause the alarm never to turn on?
6. What could cause the alarm never to turn off?
7. How could a burglar get by this system?

05 Burglar alarm

X0—Reset X1—Door X2—Window Y0—Alarm

1 X0 X1 X2 Y0 (Y0 OUT)

2 (End)

3 (NOP)

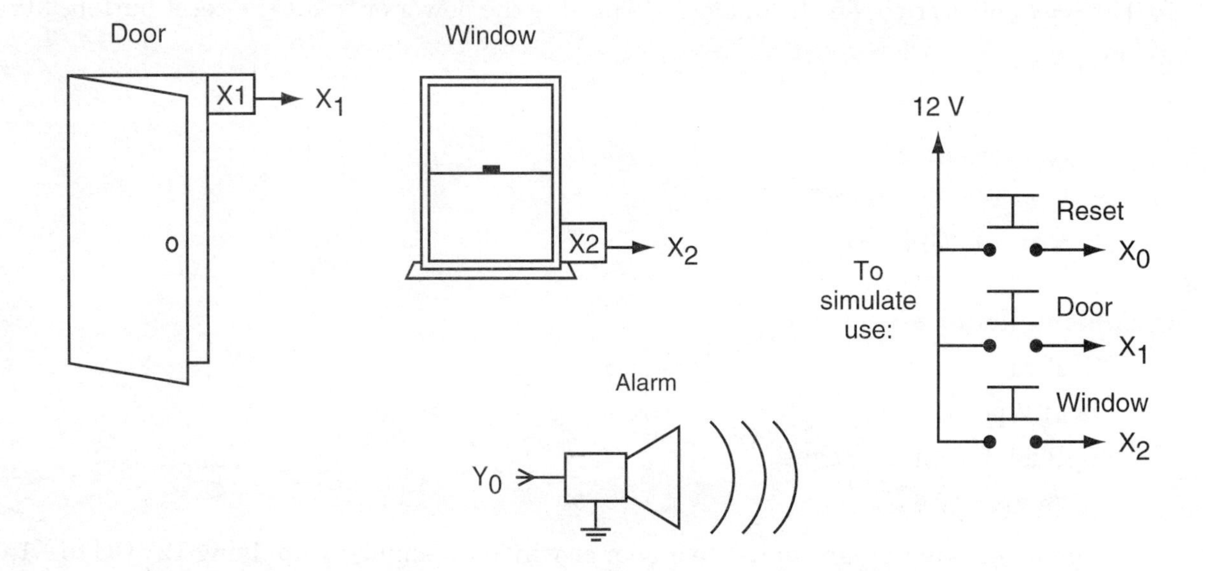

Figure 3.4

Name ______________________________ Date ______________

Lab

Microwave Oven (AND-GATE)

A microwave oven has a start switch X1, a cancel or stop switch X0, and a door safety switch X2. Refer to Figure 3.5. The oven will turn on (Y0 on) if the start and the safety switch are both closed and the stop switch is not activated.

To recap:

X0—Stop switch

X1—Start switch

X2—Safety switch

Y0—Output

Equipment Required

1. PLC
2. 12V power supply
3. LED
4. 1kohm resistor
5. Push button switches

To demonstrate the lab, apply 12VDC to X1, X2, while observing the output.

Questions

1. What happens to Y0 when X1 and X2 are activated and X0 is not activated?
2. If the microwave is on and X0 is activated, what happens to Y0?
3. If the microwave never turns on and the start switch is good, what could be wrong with X0 or X2?
4. If the microwave does not turn off when the door is open, what could be wrong?
5. If the microwave oven never turns on after activating X1 and the door is closed, what could be wrong?

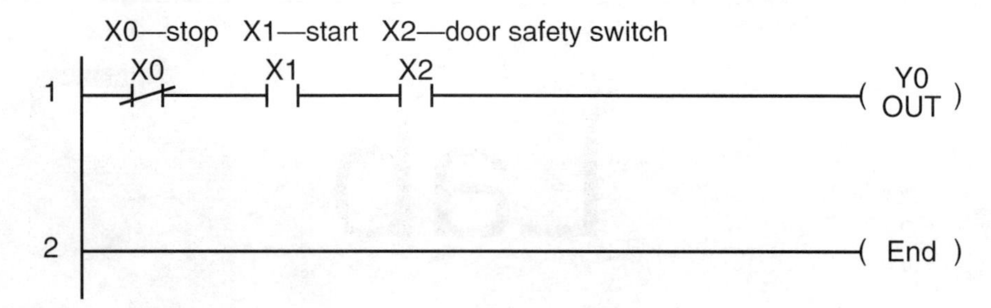
05 Microwave Oven (and Gate)
X0—stop X1—start X2—door safety switch
X0
X1
X2
1
Y0
OUT
2
End

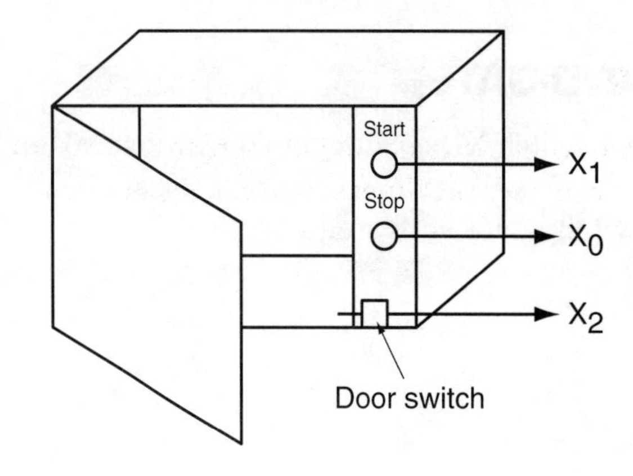
Start
Stop
X1
X0
X2
Door switch

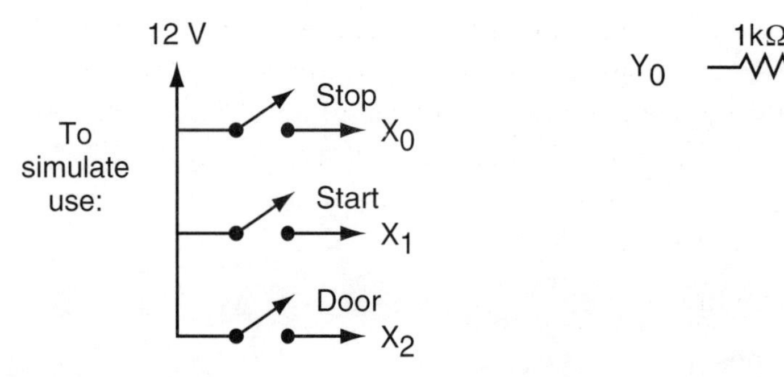
12 V
To
simulate
use:
Stop
X0
Start
X1
Door
X2
1kΩ
Y0

Figure 3.5

Name ______________________________ Date ______________

Lab

Write Date or Not?

Write a program to determine whether or not to date someone. The criteria are looks, money, family, job, and current or prior mental illness. Let the inputs be as follows:

X0—Looks

X1—Money

X2—Family

X3—Job

X4—Mental illness

Activating any of these inputs will mean they have good looks, money, a good family, a good job, a mental illness. The program should be written so that if the person has good looks **and** a good family **and** a good job, but **not** a mental illness, then the output will turn on, **or** if they have money, the output will turn on. Let the output be Y0. Connect to Y0 a 1kohm resistor and a green LED. The status of this light indicates date worthiness. Have the output Y0 turn on if they are date worthy.

Refer to Figure 3.3 for help.

Equipment Required

1. PLC
2. 12V power supply
3. LED
4. 1kohm resistor
5. Push button switches

Name ______________________________ Date ______________

Lab 

Fuel Tank (NAND-GATE)

On some large vehicles there are two fuel tanks, a main and a reserve. Fuel level sensors reporting half full or not half full measure the two tanks' levels. One sensor is connected to X0, the other to X1. The systems are designed to indicate (with a red light connected to Y0) when both tanks are less than half full using a NAND gate in the software. If both sensors put out 1 (the tank is more than half full), the NAND gate puts out 0; otherwise it puts out 1. Refer to Figure 3.6.

To recap:

X0—Fuel sensor 0

X1—Fuel sensor 1

Y0—Warning light

Eqipment Required

1. PLC
2. 12V power supply
3. LED
4. 1kohm resistor
5. Push button switches

To demonstrate the lab, alternately apply 12VDC to X0, X1, while observing the output.

Questions

1. What happens if the sensors both report 0?
2. What happens if the sensors both report 1?
3. What happens if the sensors report something different?
4. Verify that the NAND configuration will ensure correct operation by checking the logic truth table.
5. How could you test this system to ensure it will work correctly?

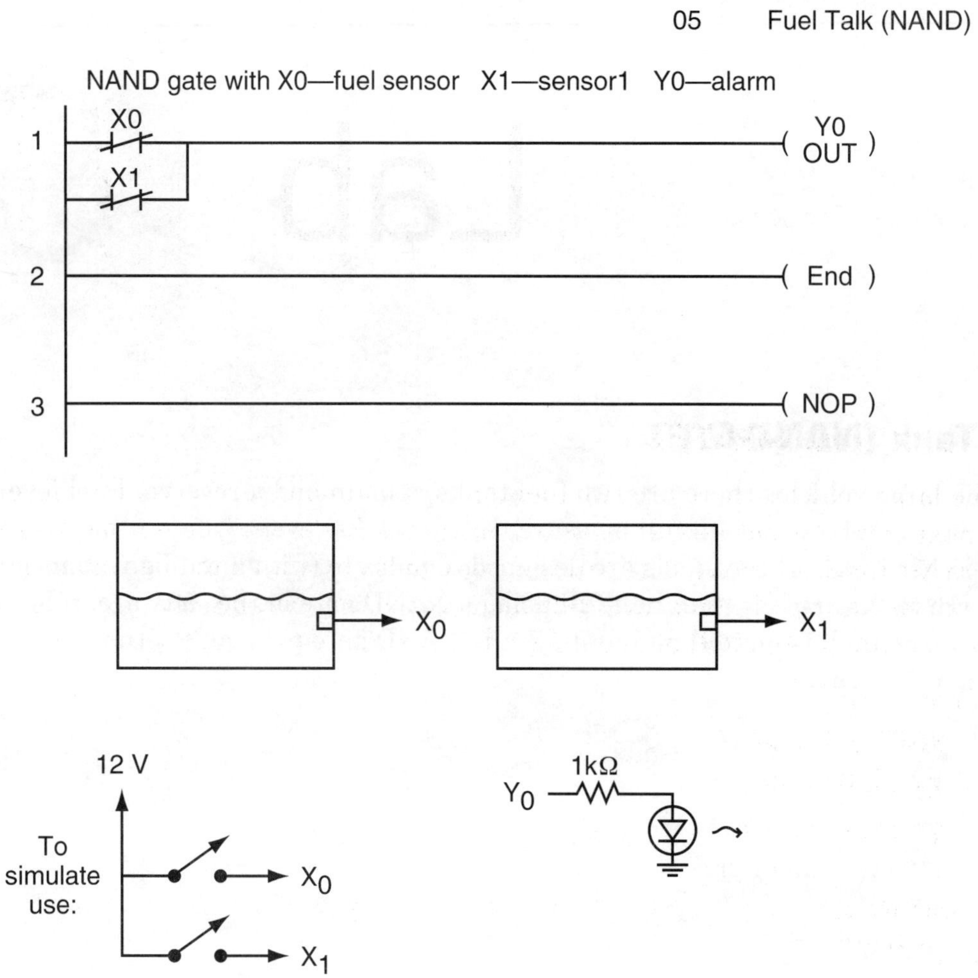

05 Fuel Talk (NAND)

NAND gate with X0—fuel sensor X1—sensor1 Y0—alarm

1 X0 X1 (Y0 OUT)

2 (End)

3 (NOP)

X_0 X_1

12 V

To simulate use: X_0 X_1

1kΩ

Y_0

Figure 3.6

Name ______________________ Date ______________

Fuel Tank (NOR-GATE)

On some large vehicles there are two fuel tanks, a main and a reserve. Fuel level sensors report near empty or not near empty measuring the two tanks' levels. One sensor is connected to X0, the other to X1. The system is designed to tell you (with a warning light connected to Y0) when both tanks are near empty using a NOR gate in the software. If both sensors measure 0, the NOR gate puts out 1, otherwise it puts out 0. Refer to Figure 3.7.

To recap:

X0—Fuel sensor 0

X1—Fuel sensor 1

Y0—Warning light

Equipment Required

1. PLC
2. 12V power supply
3. LED
4. 1kohm resistor
5. Push button switches

To demonstrate the lab, alternately apply 12VDC to X0, X1, while observing the output.

Questions

1. What happens if the sensors report 0?
2. What happens if the sensors report something different?
3. Verify that an NOR configuration will ensure correct operation by checking the logic truth table.
4. How could you test this system to ensure it will work correctly?

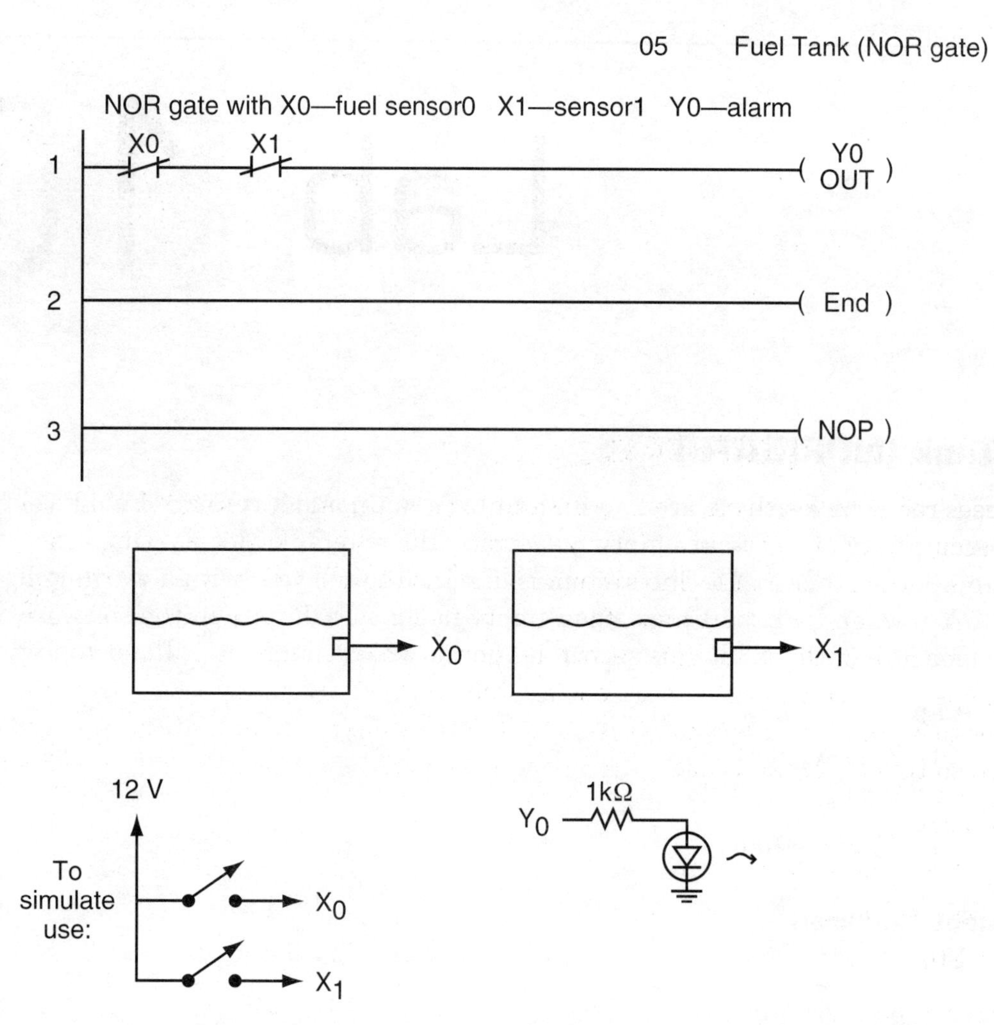
05 Fuel Tank (NOR gate)
NOR gate with X0—fuel sensor0 X1—sensor1 Y0—alarm
1
X0
X1
Y0
OUT
2
End
3
NOP
X0
X1
12 V
To
simulate
use:
X0
X1
1kΩ
Y0

Figure 3.7

Name ______________________________ Date ______________

Lab

NASA Sensors (XOR-GATE)

NASA uses redundant sensors in its spacecraft to ensure confidence that what is being measured is accurate. In other words, if the temperature of an engine is measured, there are two sensors measuring the engine. In this example, two temperature sensors are measuring the engine. The sensors put out a high if the temperature is high and put out a low if the sensors measure a low. If the sensors report a different result, a siren is sounded notifying the mission specialist that there is a potential problem. One of the sensors is connected to X0, the other to X1, and the siren to Y0. X0 and X1 are connected in a XOR configuration in the software so that if the X0 and X1 are not the same, the siren sounds. Refer to Figure 3.8.

To recap:

X0—Temperature sensor 0

X1—Temperature sensor 1

Y0—Alarm

Equipment Required

1. PLC
2. 12V power supply
3. Buzzer
4. Push button switches

To demonstrate the lab, alternately apply 12VDC to X0, X1, while observing the output.

Questions

1. What happens if the sensors report the same temperature?
2. What happens if the sensors report a different temperature?
3. Verify that an XOR configuration will ensure correct operation by checking the logic truth table.
4. How could you test this system to ensure it will work correctly?
5. How would you change the software so that the siren latches once activated?

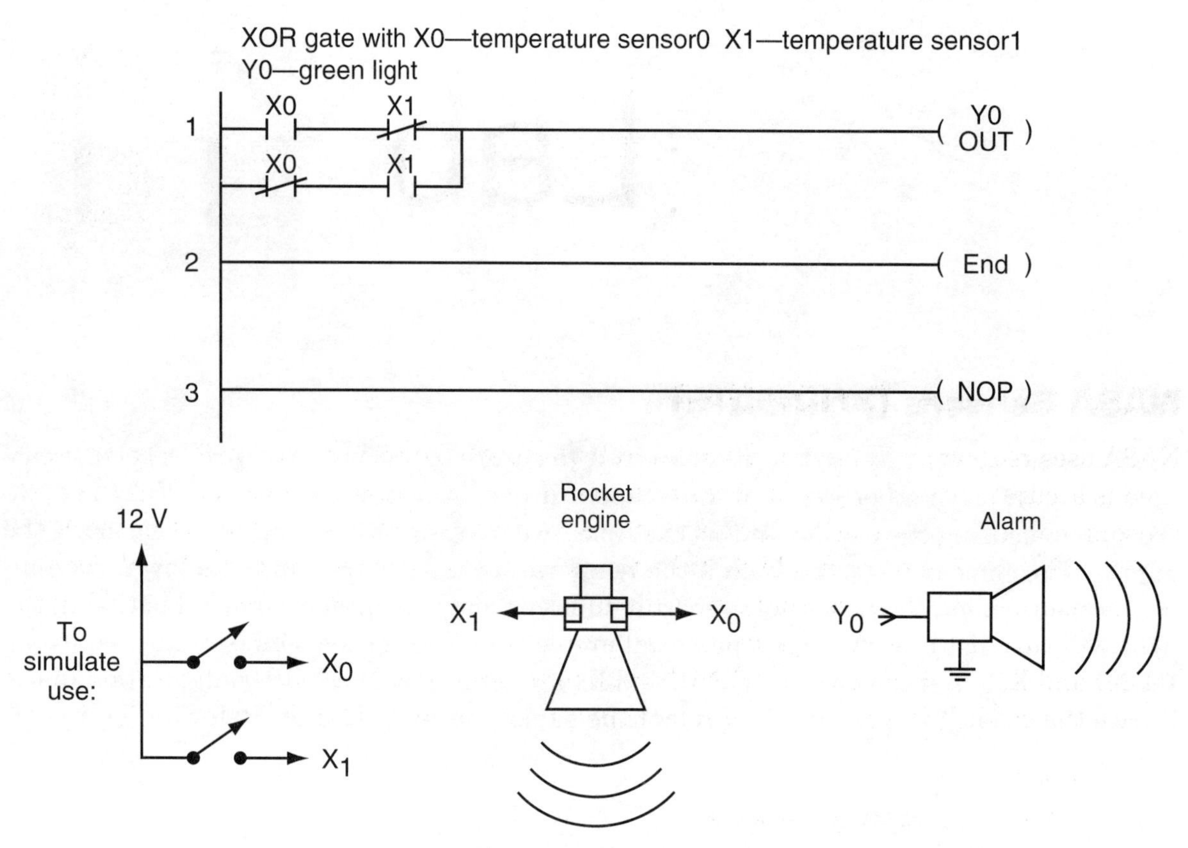
05 NASA sensors (XOR gate)
XOR gate with X0—temperature sensor0 X1—temperature sensor1
Y0—green light
X0
X1
1
Y0
OUT
X0
X1
2
End
3
NOP
12 V
To
simulate
use:
X_0
X_1
Rocket
engine
X_1
X_0
Alarm
Y_0

Figure 3.8

Name ______________________ Date ______________

NASA Sensors (XNOR-GATE)

NASA uses redundant sensors in its spacecraft to ensure confidence that what is being measured is accurate. In other words, if the temperature of an engine is measured, there are two sensors measuring the engine. In this example, two temperature sensors are measuring the engine. The sensors put out a high if the temperature is high and put out a low if the sensors measure a low. If the sensors report the same value, either both high or both low, a green light will turn on. One of the sensors is connected to X0, the other to X1, and the siren to Y0. X0 and X1 are connected in a XNOR configuration in the software so that if the X0 and X1 are the same, the green light will be activated. Refer to Figure 3.9.

To recap:

X0—Temperature sensor 0

X1—Temperature sensor 1

Y0—LED

Equipment Required

1. PLC
2. 12V power supply
3. LED
4. 1kohm resistor
5. Push button switches

To demonstrate the lab, alternately apply 12VDC to X0, X1, while observing the output.

Questions

1. What happens if the sensors report the same temperature?
2. What happens if the sensors report a different temperature?
3. Verify that an XNOR configuration will ensure correct operation by checking the logic truth table.
4. How could you change the software so that the light latches on once activated?
5. How could you test this system to ensure it will work correctly?

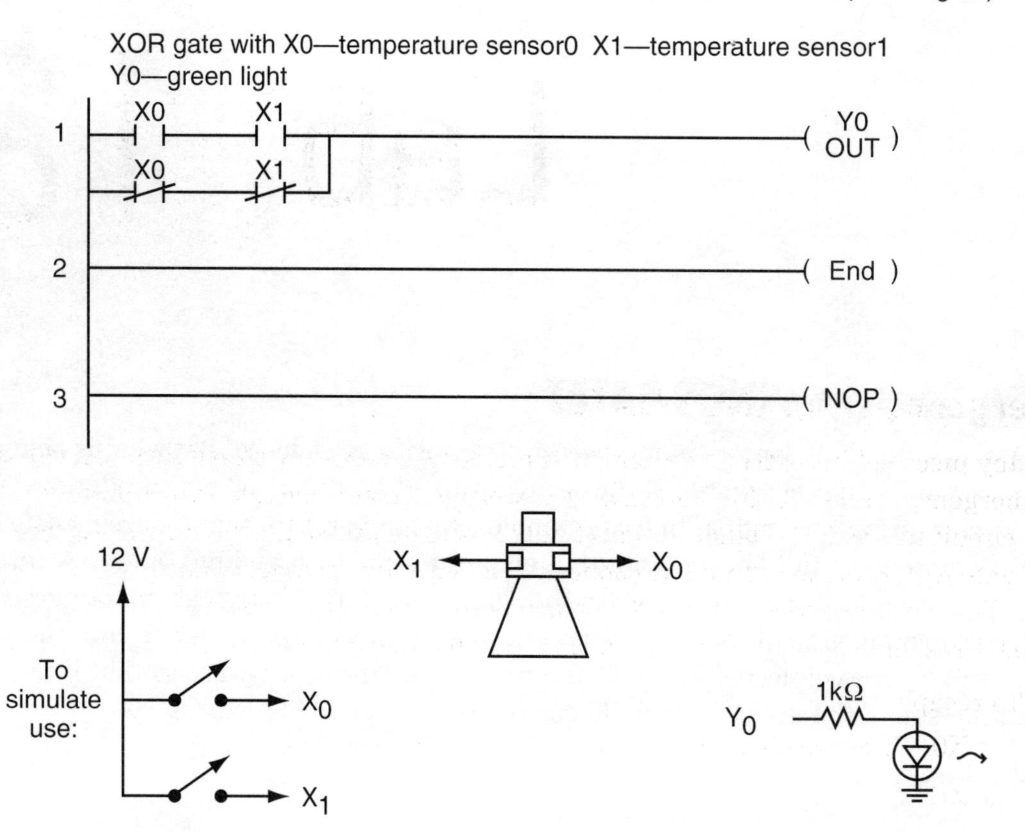

05 NASA sensors (XNOR gate)
XOR gate with X0—temperature sensor0 X1—temperature sensor1
Y0—green light
1
X0
X1
X0
X1
Y0
OUT
2
End
3
NOP
12 V
To
simulate
use:
X0
X1
X1
X0
1kΩ
Y0

Figure 3.9

Name ______________________ Date ____________

Lab 13

Emergency Stop (NOT-GATE)

In many pieces of industrial equipment there is an emergency stop button. In the event of an emergency, pressing this button will power off the equipment. Typically it is used in a latch circuit to break the latch so that it only has to be pressed momentarily. In this lab, the NOT gate will be demonstrated by connecting a normally open push button (N.O.P.B.) switch to X0. The output of the circuit is Y0, which is connected to a 1kohm resistor in series with a LED. Pressing the switch will turn off the light. Refer to Figure 3.10.

To recap:

X0—Emergency stop

Y0—LED

Equipment Required

1. PLC
2. 12V power supply
3. LED
4. 1kohm resistor
5. Push button switch

To demonstrate the lab, alternately apply 12VDC to X0 while observing the output.

Questions

1. Is the LED on or off when the switch SW0 is pressed?
2. Is the LED on or off when the switch SW0 is not pressed?
3. If the SW0 is open, is X0 in the program open or closed?
4. If SW0 is closed, is X0 in the program open or closed?

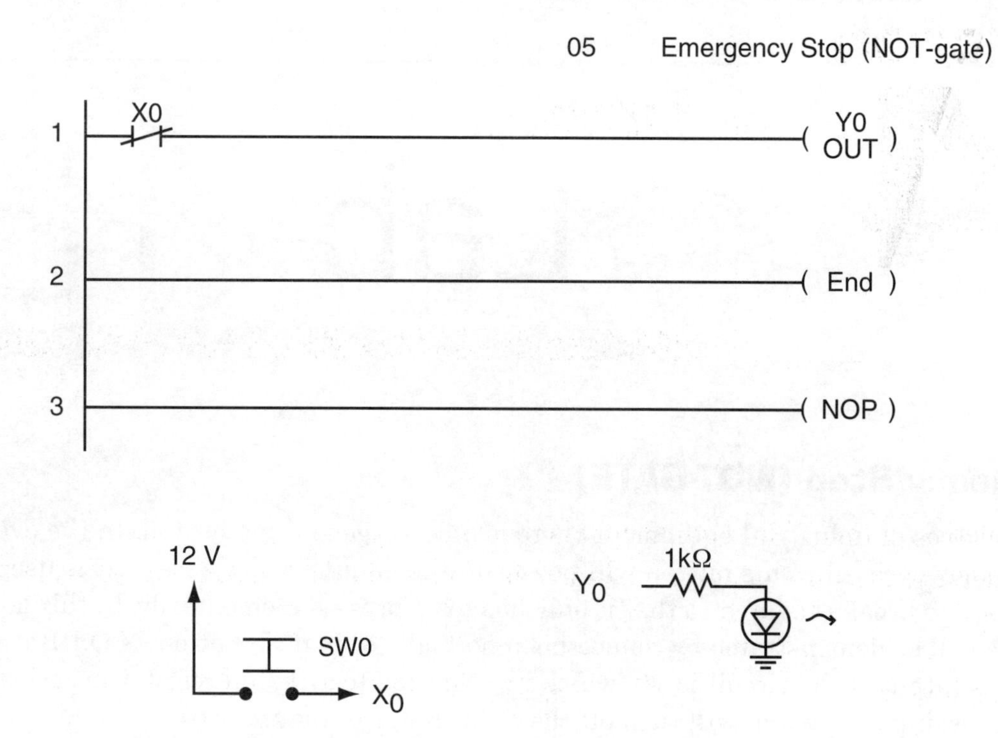
05 Emergency Stop (NOT-gate)
1
X0
Y0
OUT
2
End
3
NOP
12 V
SW0
X_0
1kΩ
Y_0

Figure 3.10

SECTION 4

Timers

Objective

Upon completion of these labs, you should be able to:

- Understand the use of delay on, immediate on, and accumulating timers in ladder logic.

Introduction

Timers can record how long an input has been activated, or delay or limit the length of time an output is turned on. The timers used in this manual are all "delay on." This means that if the timer is activated, its output will turn on after a period of time has elapsed (like an alarm clock). "Immediate on" timers' output turns on immediately after activation and then turns off after some designated time. We will use two types of timer functions in this manual, timer "TMR" and timer accumulate "TMRA." Illustrated in Figures 4.1 and 4.2 is TMR.

Figure 4.1 illustrates a delay on timer. The timer function TMR has an input X0 which activates the timer. As long as it is activated, the timer is running. If the timer is deactivated, the timer will reset to 0. The timer count is determined by K50. K50 means 5 seconds. After 5 seconds of activation, the timer output will turn on. The timer output is T0. When the count is reached, T0 will turn on, closing switch T0, and output Y0 will turn on.

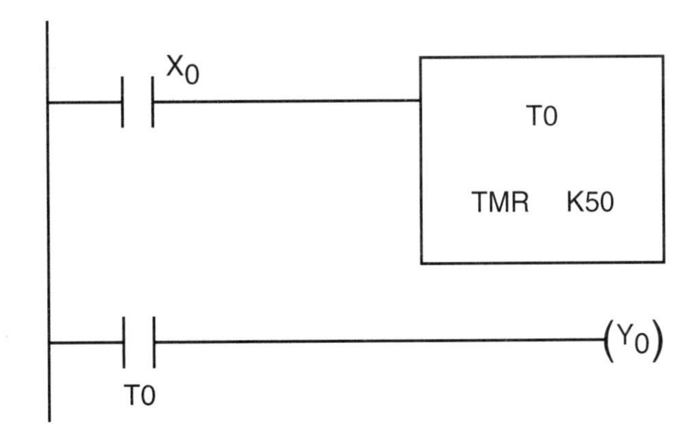

Figure 4.1 A delay timer. After 5s of X0 being activated, Y0 will turn on.

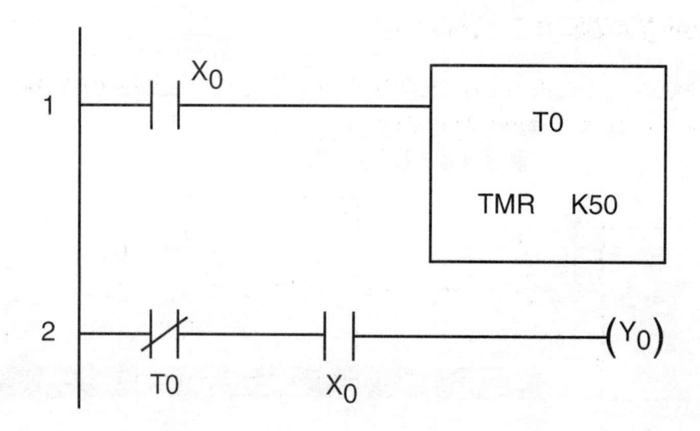

Figure 4.2 An immediate on timer. When the timer reaches its count 5s, the normally closed switch T0 will open and turn off Y0.

Figure 4.2 depicts an immediate on timer. On rung 2, T0 is normally closed and X0 is normally open. When X0 is closed, output Y0 will turn on immediately because switch T0 and X0 in the second rung are closed. When the timer reaches its count, the normally closed switch T0 will open and turn off Y0.

Comparative Contacts (LABS 17-24, 27-28)

Besides normally open and normally closed switches or contacts, PLCs have contacts that compare one side of the contact to the other. For example, look at Figures 4.3–4.6.

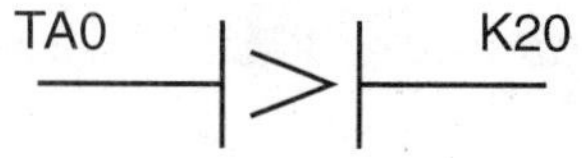

TA0 ⊣>⊢ K20

Figure 4.3 If the time from timer T0 is greater than 2s, then this contact is closed.

TA0 ⊣<⊢ K20

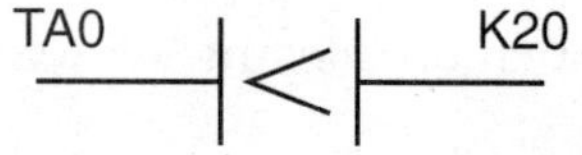

Figure 4.4 If the time from timer T0 is less than 2s, then this contact is closed.

TA0 ⊣=⊢ K20

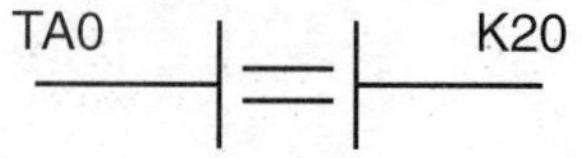

Figure 4.5 If the time from timer T0 is equal to 2s, then this contact is closed.

TA0 ⊣≠⊢ K20

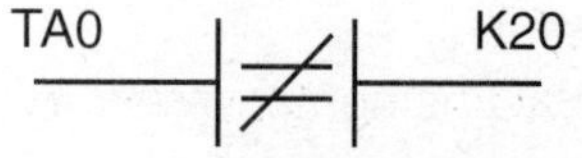

Figure 4.6 If the time from timer T0 is not equal to 2s, then this contact is closed.

Resetting the Timer (LABS 17, 18, 24)

To have a timer time out and begin again on its own, connect the timer output to its input with a normally closed contact. See Figure 4.7.

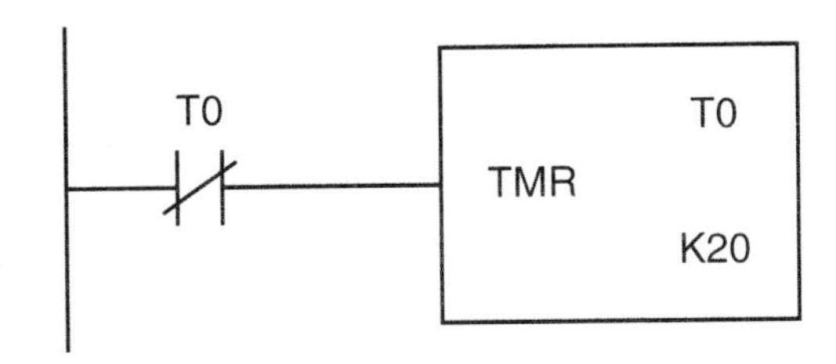

Figure 4.7 When the timer reaches its limit count, in this case 2s, the normally closed T0 contact will briefly open and reset the timer. The normally closed T0 will then close and the timer will begin again.

Timer Accumulate (TMRA) (LABS 25-28)

When timer TMR's input opens, the timer resets to 0s. Comparing this to TMRA, when TMRA's input opens, the time is kept. If the input is closed again, the time starts to accumulate from where it left off. The other difference between TMR and TMRA is that TMRA has a reset input, zeroing out the clock, Figure 4.8.

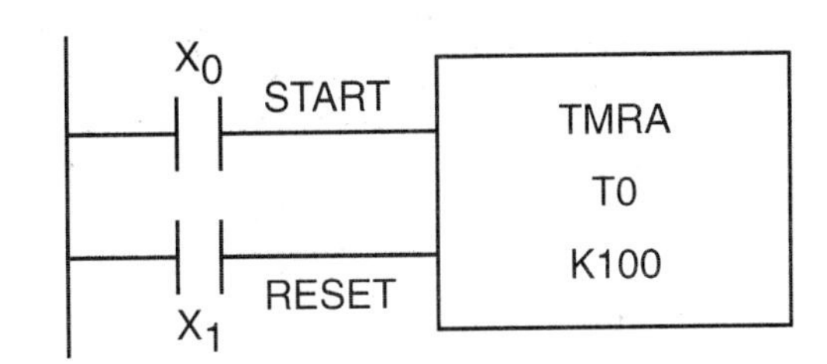

Figure 4.8 The TMRA function records how long the input was closed. If the input is opened, the time is retained. The reset will zero out the clock.

Name ______________________ Date ____________

Lab 14

Delay On, Immediate On Light

Two lights are controlled by one switch X0. One light, connected to Y1, turns on immediately after X0 is pressed and stays on for 5s; the other, connected to Y0, turns on 5s after X0 is activated and stays on. Refer to Figure 4.9.

To recap:

X0—Activate light

Y0—Delay on light

Y1—Immediate on light

Equipment Required

1. PLC
2. 12V power supply
3. LEDs
4. 1kohm resistors
5. Push button switches

To demonstrate the lab, alternately apply 12VDC to X0 while observing the outputs.

Questions

1. What happens when X0 is activated?
2. What happens to output Y0 after 5s of activating X0?
3. Does the normally closed switch T0 open after the timer has been on for 5s?
4. What is K50 in the program?
5. What happens to output Y1 after 5s of activating X0?

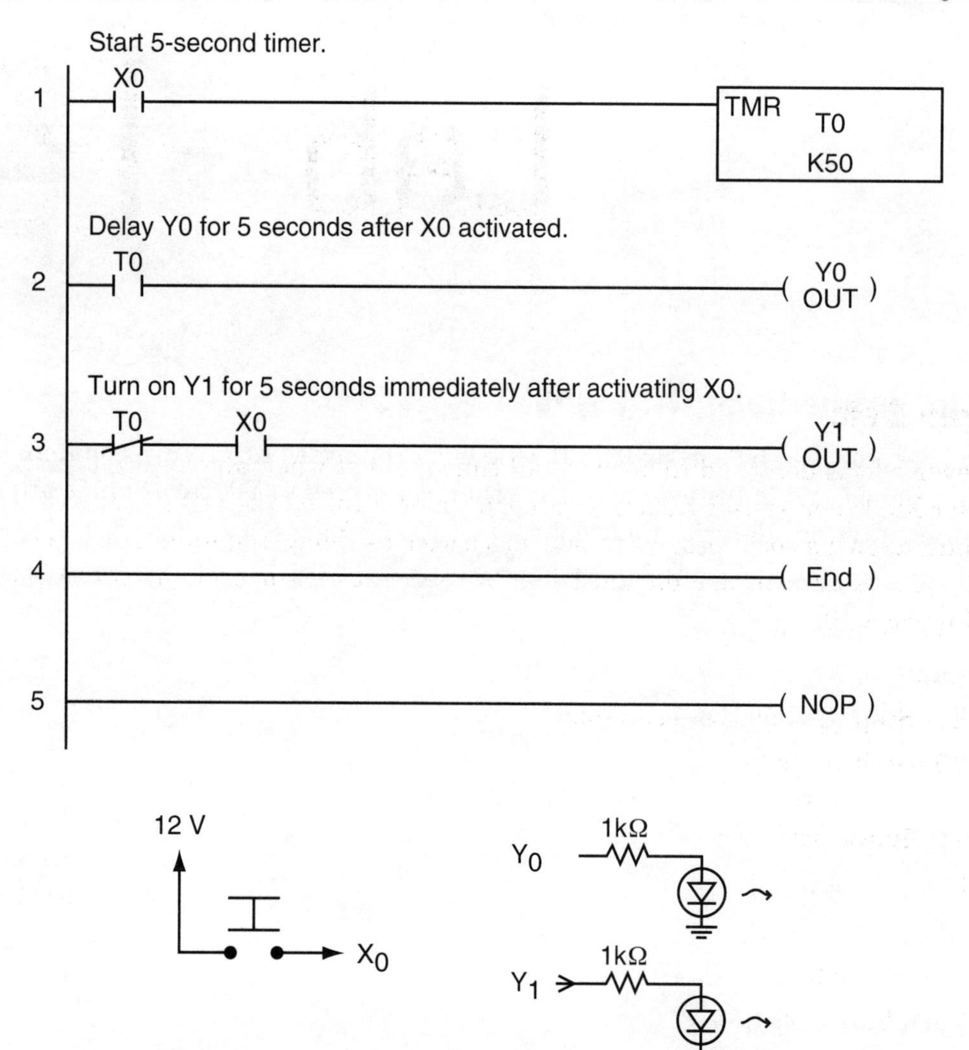
05 Delay-on/immediate-on light
Start 5-second timer.
1
X0
TMR
T0
K50
Delay Y0 for 5 seconds after X0 activated.
2
T0
Y0
OUT
Turn on Y1 for 5 seconds immediately after activating X0.
3
T0
X0
Y1
OUT
4
End
5
NOP
12 V
X0
Y0
1kΩ
Y1
1kΩ

Figure 4.9

Name ______________________________ Date ______________

Lab 15

Conveyer Belt

The conveyer belt is designed to be on at all times except when an object on the belt activates sensor X0. When sensor X0 is activated, a timer turns off the belt (through output Y0) for 5 seconds, allowing some process to take place such as filling a container on the belt. After 5 seconds, the belt continues on until it hits the next sensor and the process repeats. Refer to Figure 4.10.

To recap:

X— Stop the belt for 5 seconds

Y0—Belt power

Equipment Required

1. PLC
2. 12V power supply
3. 12VDC motor
4. Push button switch

To demonstrate the lab, apply 12VDC to X0 and hold it there for at least 5s while observing the output.

Questions

1. What happens if X0 is activated?
2. What happens if X0 is not activated?
3. If there are three X0 sensors on the belt, how many times will the belt stop?
4. If the belt never stops, what could be wrong with the system?
5. What turns off the belt?
6. If the belt stops and never turns on again, what could be the problem?
7. What type of sensor could be used to stop the belt?

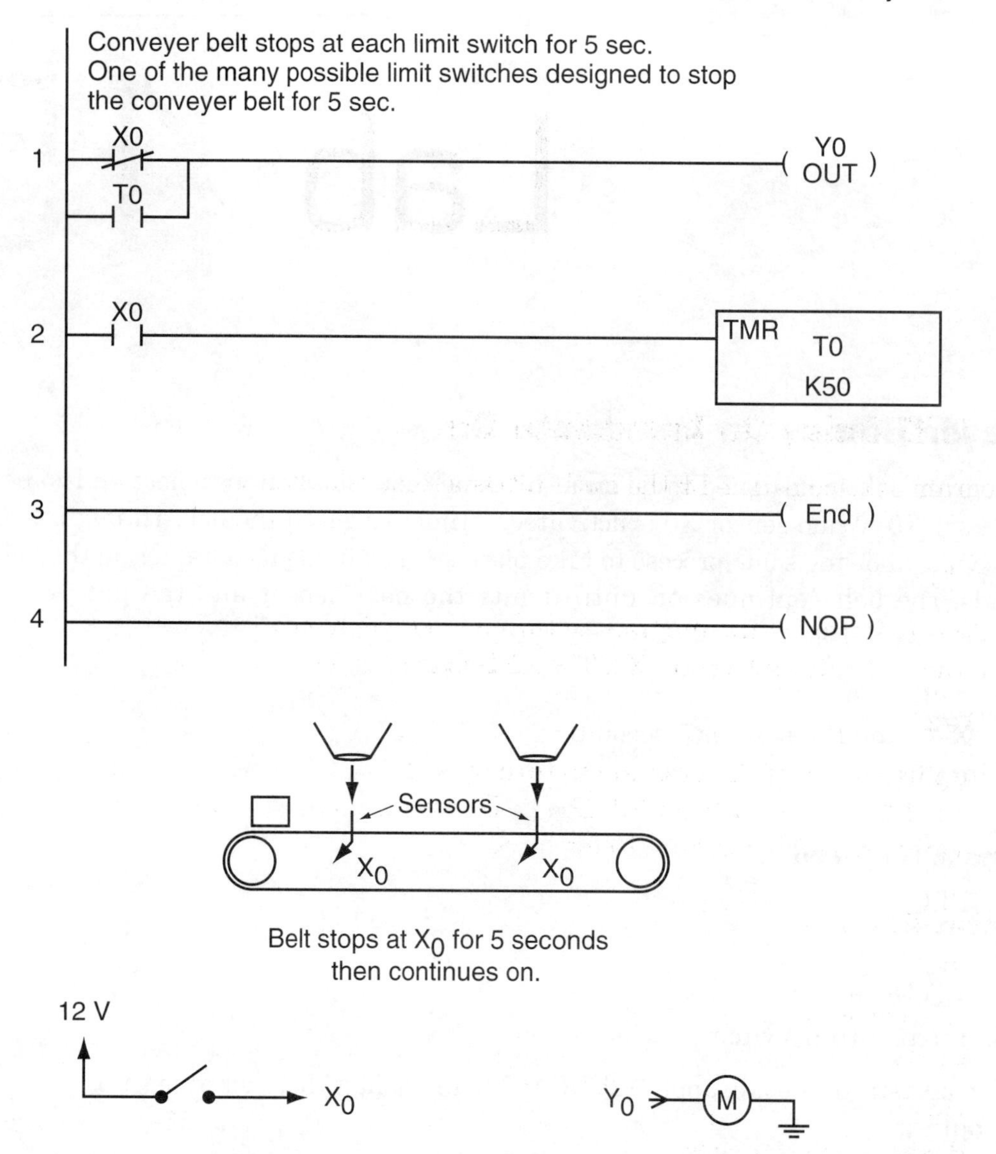
05 Conveyer Belt
Conveyer belt stops at each limit switch for 5 sec.
One of the many possible limit switches designed to stop the conveyer belt for 5 sec.
1
X0
T0
Y0
OUT
2
X0
TMR
T0
K50
3
End
4
NOP
Sensors
X0
X0
Belt stops at X0 for 5 seconds then continues on.
12 V
X0
Y0
M

Figure 4.10

Name ____________________ Date ____________

Lab

Write LED Delay On Immediate On

This program is to control 2 LEDs. Both LEDs will be controlled by the one program you write.

LED #1

Write this section of the program to turn on and latch on a LED with an input X0 and output to the LED Y0, and a reset X1. The LED is to turn on 4 seconds after X0 is activated.

LED #2

Write this section of the program to turn on and latch on a LED with an input X2 and output to the LED Y1, and a reset X3. The LED is to turn on immediately for 4 seconds after X0 is activated. Refer to Figure 4.10 for help.

Equipment Required

1. PLC
2. 12V power supply
3. LEDs
4. 1kohm resistors
5. Push button switches

Name ______________________________ Date ______________

Lab

Traffic Light

The green light on a traffic light is controlled by output Y0 and is on for 3 seconds, the yellow light by Y1 and on for 3s, and the red light by Y2 and on for 4s. The timer is self-starting and self-resetting by T0. Refer to Figure 4.11.

To recap:

Y0—Green light

Y1—Yellow light

Y2—Red light

To demonstrate the lab, run the program and observe the time sequence of the outputs.

Equipment Required

1. PLC
2. 12V power supply
3. LEDs
4. 1kohm resistors

Questions

1. Draw a timing diagram showing when the green, yellow, and red lights are on.
2. If the yellow light never turns on, what could the problem be?
3. What causes the timer to reset?
4. In the timer function, how many seconds is K100?
5. What keeps the green light from staying on for more than 3s?

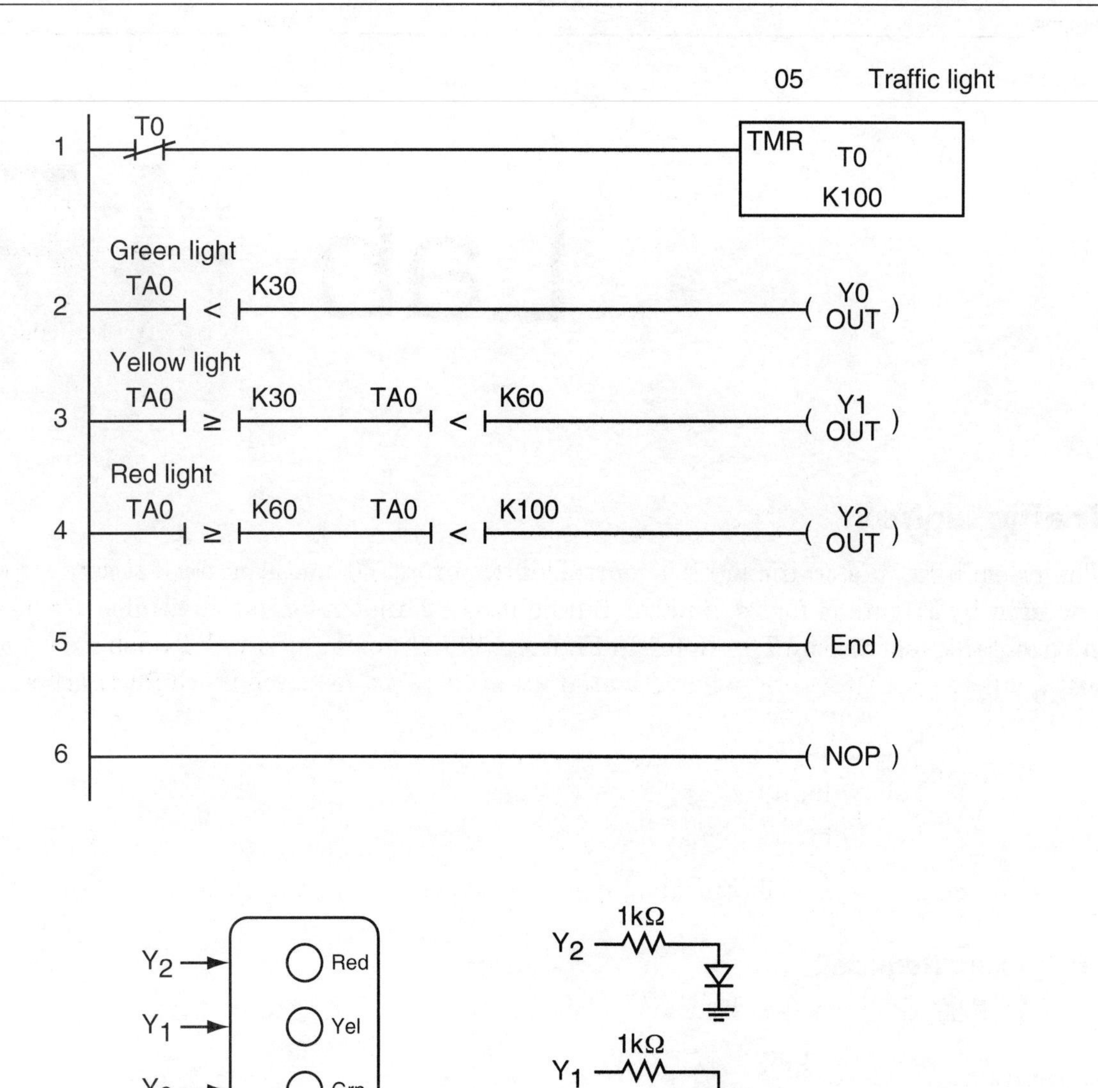
05 Traffic light
1
T0
TMR
T0
K100
Green light
2
TA0
<
K30
Y0
OUT
Yellow light
3
TA0
≥
K30
TA0
<
K60
Y1
OUT
Red light
4
TA0
≥
K60
TA0
<
K100
Y2
OUT
5
End
6
NOP
Y2
Y1
Y0
Red
Yel
Grn
1kΩ
Y2
1kΩ
Y1
1kΩ
Y0

Figure 4.11

Name ______________________ Date ____________

Traffic + Walk Light

The green light on a traffic light is controlled by output Y2 and is on for 6 seconds, the yellow light by Y3 and is on for 3s, and the red light by Y4 and is on for 3s. The timer is self-starting and self-resetting by T0. A flashing walk light connected to Y1 flashes at a rate of one flash every half second when a walk button X0 is pressed and the light is green. Refer to Figure 4.12.

To recap:

Y1—Flashing walk light
Y2—Green light
Y3—Yellow light
Y4—Red light
X0—Walk light

Equipment Required

1. PLC
2. 12V power supply
3. LEDs
4. 1kohm resistors
5. Momentary switch

To demonstrate the lab, run the program and observe the time sequence of the outputs. Momentarily apply 12VDC to X0 and observe the walk light Y1.

Questions

1. Draw a timing diagram showing when the green, walk light, yellow, and red light are on.
2. If the yellow light never comes on, what could the problem be?
3. What causes timer T0 to reset?
4. In the timer function, how many seconds is K120?
5. What keeps the green light from staying on for more than 6s?

6. What keeps the flashing walk light from flashing at times other than when the green light is on?

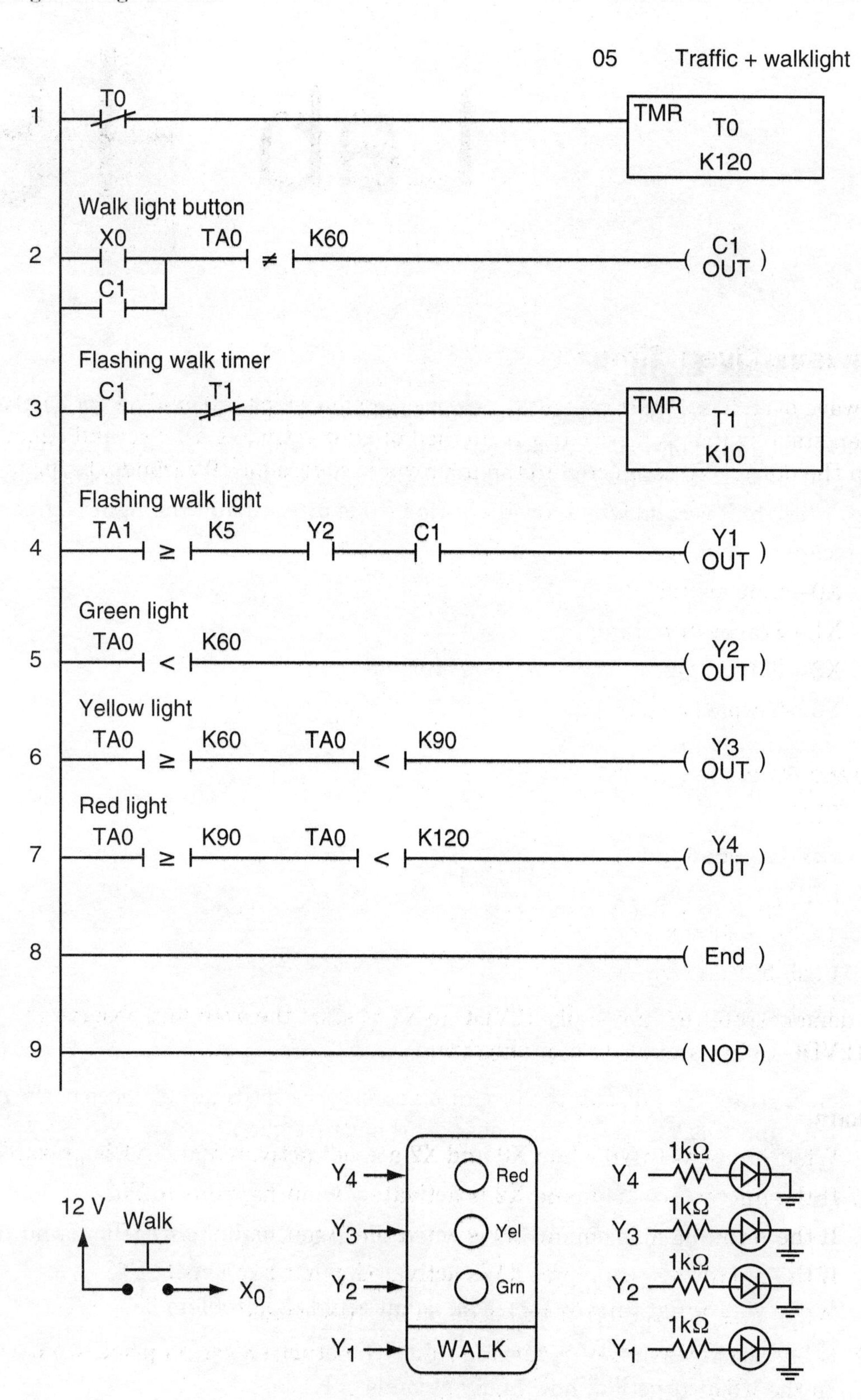

Figure 4.12

Name ______________________________ Date ____________

Microwave Oven Timer

A microwave oven has a start switch X1, a cancel or stop switch X0, and a door safety switch X2. When start switch is pressed, it is latched on and activates a 10 second timer, which turns on the output Y0 connected to the microwave source for 10 seconds. Refer to Figure 4.13.

To recap:

X0—Stop switch
X1—Start switch
X2—Safety switch
Y0—Output

Equipment Required

1. PLC
2. 12V power supply
3. LED
4. 1kohm resistor
5. Push button switches

To demonstrate the lab, apply 12VDC to X1 to start the oven and observe the output. Apply 12VDC to X0, and X2 to stop the oven.

Questions

1. What happens to Y0 when X0 and X2 are not activated and X1 is pressed?
2. If the microwave is on and X2 is activated, what happens to Y0?
3. If the microwave is on and X0 is activated, what happens to Y0?
4. If the microwave is on and X1 is activated, what happens to Y0?
5. What does Y0 do after it has been on for 10 seconds?
6. If the microwave never turns on and the start switch is good, what could be wrong with X0 or X2?

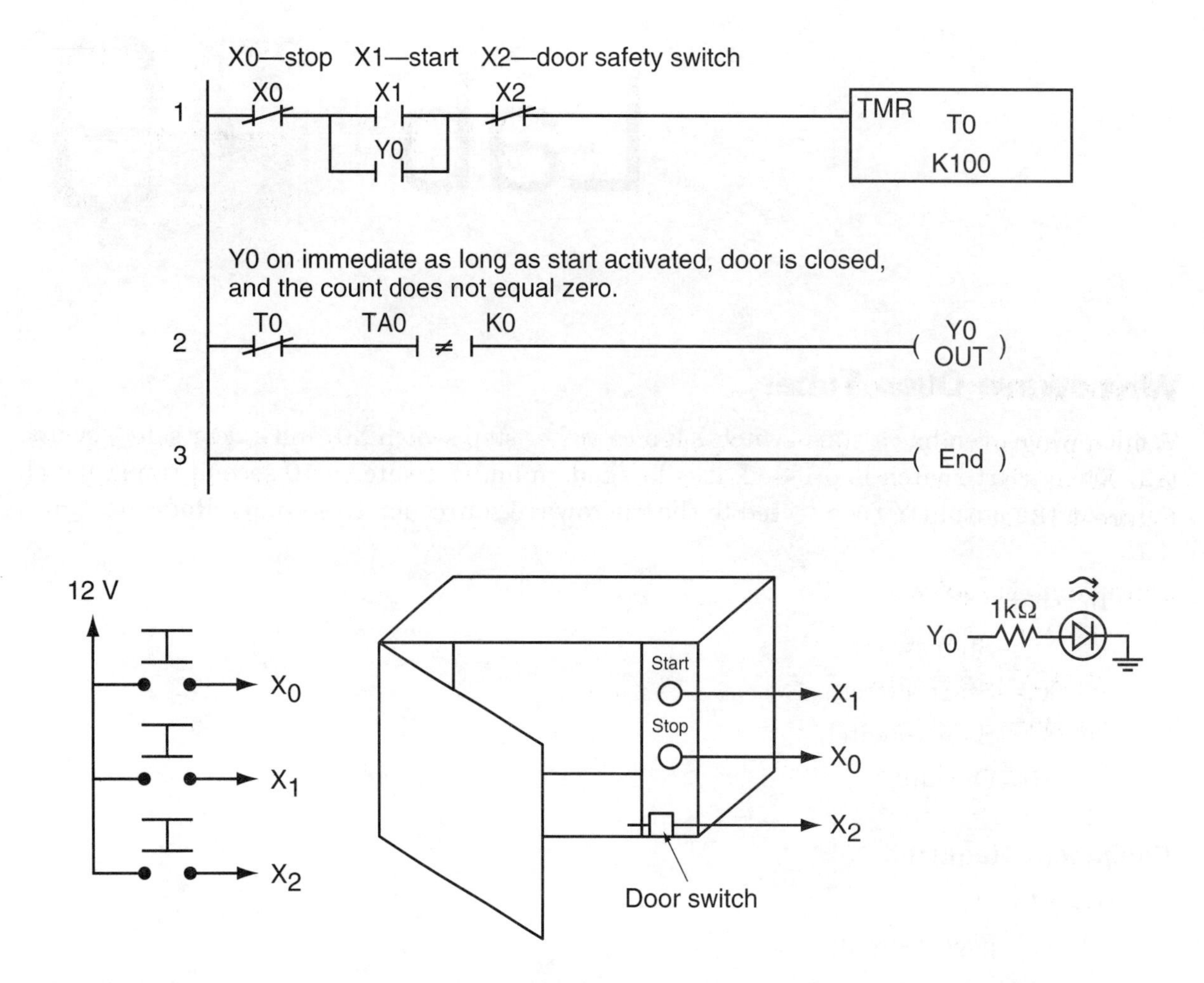
05 Microwave Oven
X0—stop X1—start X2—door safety switch
1
X0
X1
X2
Y0
TMR
T0
K100
Y0 on immediate as long as start activated, door is closed, and the count does not equal zero.
2
T0
TA0
≠
K0
Y0
OUT
3
End
12 V
X0
X1
X2
Start
Stop
X1
X0
X2
Door switch
Y0
1kΩ

Figure 4.13

Name ______________________________ Date ______________

Lab

Write Alarm Clock

Write a program to control an alarm clock that rings 8 seconds after it has been activated. Use X0 as the start button, X1 as reset, and Y0 as the output to the ringer. Refer to Figure 4.13 for help.

Equipment Required

1. PLC
2. 12V power supply
3. Buzzer
4. Push button switches

Name ______________________ Date ______________

Lab

Washing Machine

A washing machine runs through the following cycle: FILL, AGITATE, DRAIN, and then SPIN. The program begins when the start button, X0, is pressed. The machine begins to fill by opening the fill valve for 2 seconds, then agitates for 2 seconds, then drains for 2 seconds, and finally spins for 2s with the drain remaining open while spinning. Refer to Figure 4.14.

To recap:

X0—Start
Y0—Fill valve
Y1—Agitate
Y2—Drain valve
Y3—Spin

Equipment Required

1. PLC
2. 12V power supply
3. LEDs
4. 1kohm resistors
5. Push button switch

To demonstrate the lab, momentarily apply 12VDC to X0, and observe the outputs.

Questions

1. What happens to C0 when X0 is pressed?
2. What happens during the 2s to 4s period of the cycle?
3. If the machine never fills and the program is working fine, what could be the problem?
4. If the machine never drains and the program is working fine, what could be the problem?
5. If the machine never spins and the program is working fine, what could be the problem?
6. What turns off the timer in the ladder logic diagram?

7. For how long is the drain open?
8. How would you distinguish between a problem with the machine's motor and a problem with the PLC?

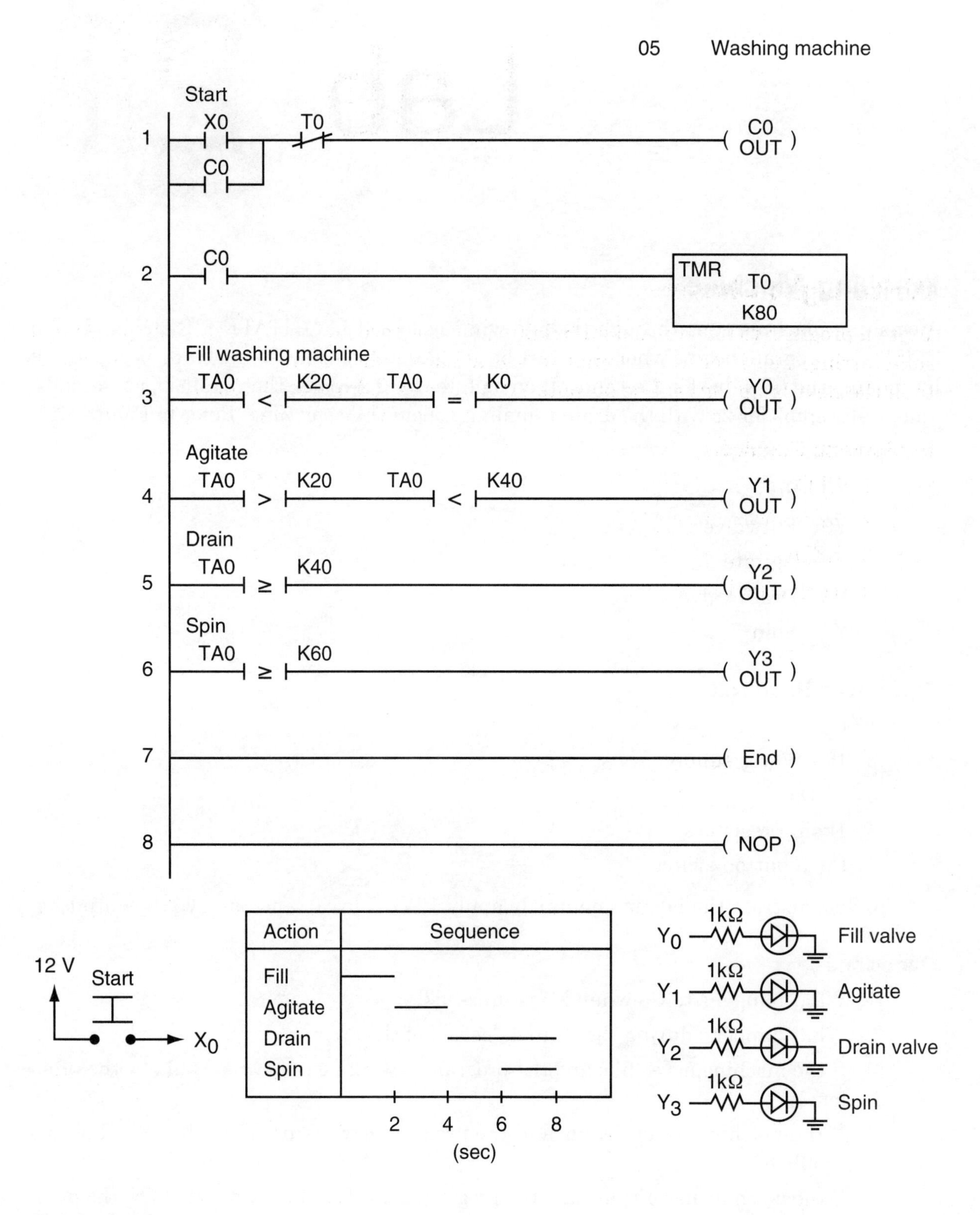

Figure 4.14

Name ______________________________ Date ______________

Lab

Write Light Chaser

Write a program to turn on and off 6 lights in a sequence and then repeat. The program is self-starting, i.e., it needs no input to start. Each light should turn on for 0.5 seconds then the next lamp lights. Use outputs Y0–Y5. Refer to Figure 4.14 for help.

Equipment Required

1. PLC
2. 12V power supply
3. LEDs
4. 1kohm resistors

Name ______________________________ Date ______________

Lab 

Mixer

The mixer software is designed to control a fluid mixer. When the start button X0 is pushed, a fill valve is opened, controlled by Y0, filling the tank. When the tank is full, X1, the full limit switch, is activated and turns off Y0. The timer is then turned on by the full switch X1 and the mixing motor is turned on for 3 seconds. After 3 seconds the mixing motor is turned off and the drain valve is opened, controlled by output Y3. The open valve stays open until the empty switch X2 is activated, turning off Y3 and hence closing the drain valve. Refer to Figure 4.15.

To recap:

X0—Start

X1—Full

X2—Empty

Y0—Open fill valve

Y2—Mixer motor power

Y3—Open drain valve

Equipment Required

1. PLC
2. 12V power supply
3. LEDs
4. 1kohm resistors
5. Push button switches

To demonstrate the lab, momentarily apply 12VDC to X0, and observe the output. To simulate the tank is full, apply and keep 12V to X1. Y0 should turn off and Y2 turn on for 3s. At this point Y3 should turn on. The tank is now draining. Remove 12V from X1 to simulate the level is dropping. To simulate the tank is empty, apply 12V to X2.

Questions

1. If X0 is activated, what happens?
2. If the float is down, are X1 and X2 activated or not?

3. If the float is up, are X1 and X2 activated or not?
4. For how long is the motor on?
5. What turns off the motor?
6. If the fill valve does not open, what is the problem?
7. If the drain valve does not open, what is the problem?
8. What activates the drain to close?
9. What activates the mixing motor to turn on?

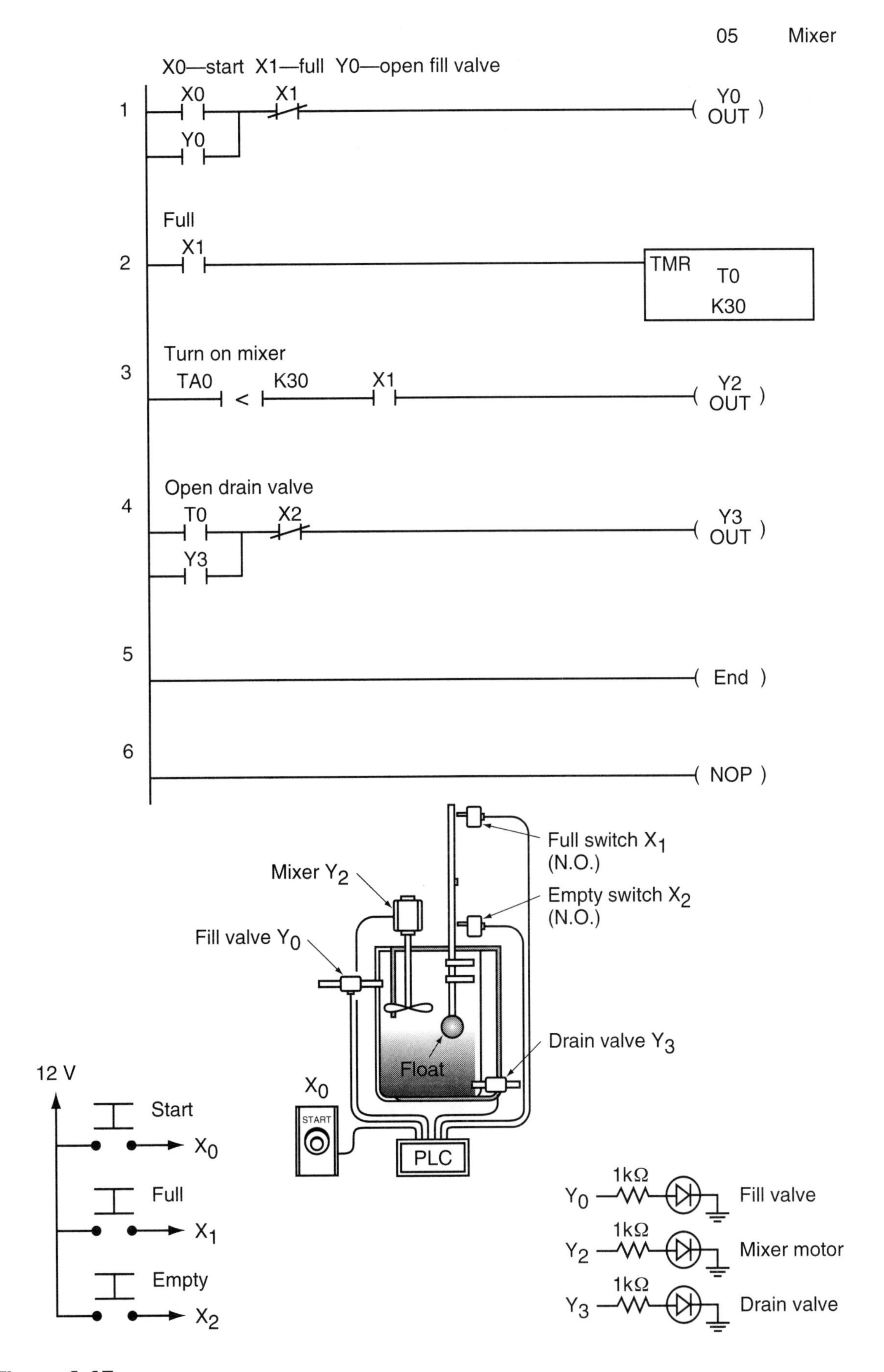
05 Mixer
X0—start X1—full Y0—open fill valve
1
X0
X1
Y0
Y0
OUT
Full
2
X1
TMR
T0
K30
Turn on mixer
3
TA0
<
K30
X1
Y2
OUT
Open drain valve
4
T0
X2
Y3
Y3
OUT
5
End
6
NOP
Full switch X1
(N.O.)
Mixer Y2
Empty switch X2
(N.O.)
Fill valve Y0
Drain valve Y3
Float
X0
START
PLC
12 V
Start
X0
Full
X1
Empty
X2
1kΩ
Y0
Fill valve
1kΩ
Y2
Mixer motor
1kΩ
Y3
Drain valve

Figure 4.15

Name ______________________________ Date ______________

Lab 

Intermittent Windshield Wipers

An intermittent windshield wiper is designed to turn the wipers on for various intervals: fast—on for 2s, off for 1s; medium—on for 2s, off for 2s; slow—on for 2s, off for 4s. A rotary switch selects the speed. Activating X0 selects the slow speed, activating X1 selects the medium speed, and activating X2 selects the fast speed. Y0 is connected to the wiper motor. Refer to Figure 4.16.

To recap:

X0—Slow

X1—Medium

X2—Fast

Y0—Wiper power

Equipment Required

1. PLC
2. 12V power supply
3. 12VDC motor
4. Rotary switch

To demonstrate the lab, alternately apply and hold 12VDC to X0, X1, X2, and observe the output.

Questions

1. What happens if X0 is selected?
2. What happens if X1 is selected?
3. What happens if X2 is selected?
4. How do the timers reset after their interval is up?
5. What could be the problem if the medium speed does not work?
6. What could be the problem if the fast speed does not work?
7. If the wiper does not turn on for any speed selection, what is probably the problem?

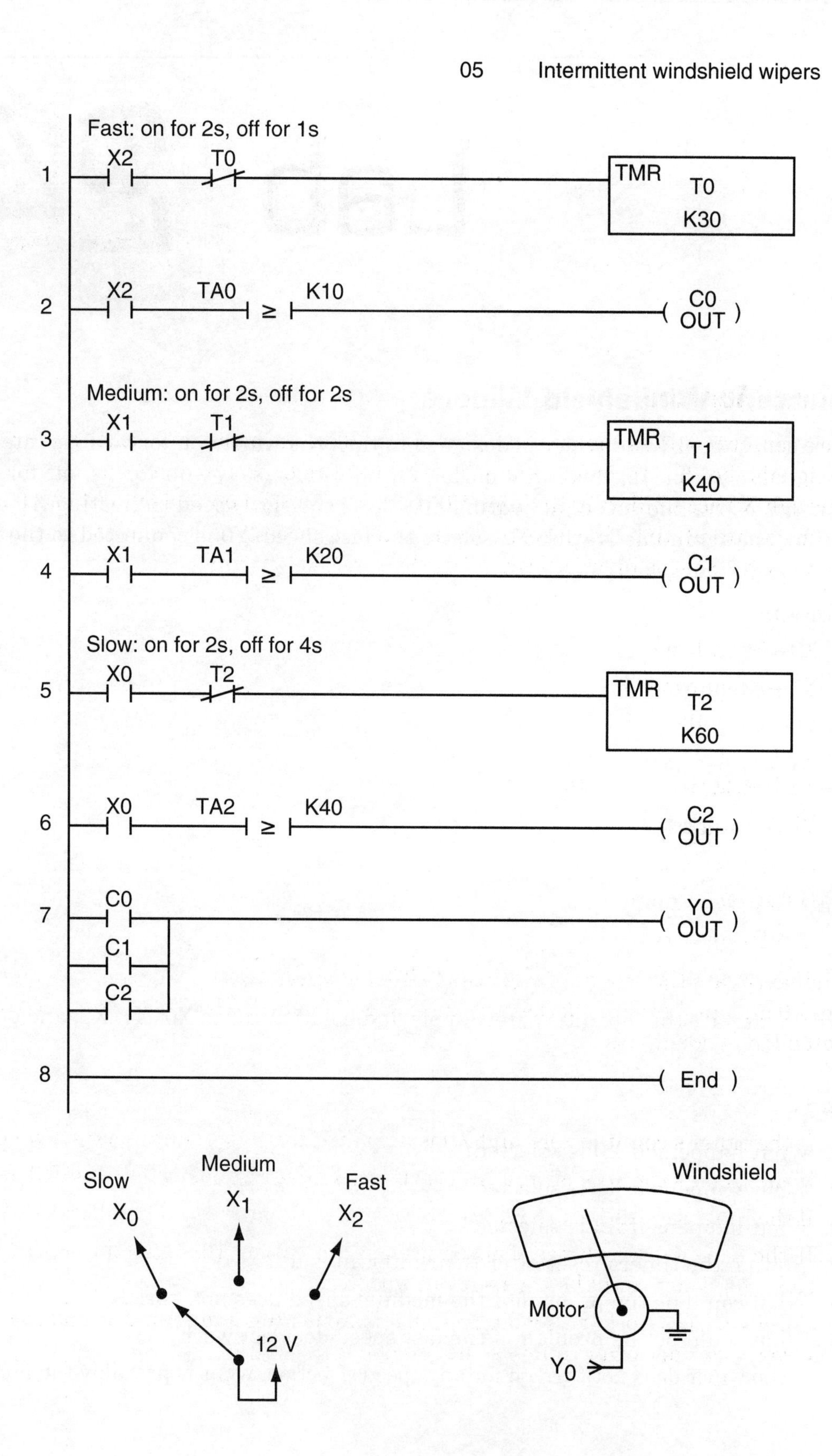
05 Intermittent windshield wipers
Fast: on for 2s, off for 1s
1
X2
T0
TMR
T0
K30
2
X2
TA0
≥
K10
C0
OUT
Medium: on for 2s, off for 2s
3
X1
T1
TMR
T1
K40
4
X1
TA1
≥
K20
C1
OUT
Slow: on for 2s, off for 4s
5
X0
T2
TMR
T2
K60
6
X0
TA2
≥
K40
C2
OUT
7
C0
C1
C2
Y0
OUT
8
End
Slow
Medium
Fast
X0
X1
X2
12 V
Windshield
Motor
Y0

Figure 4.16

Name ______________________ Date ____________

Lab 

Pressure Monitor

A pressure sensor monitors the pressure in a tank. When the pressure reaches an upper set point, it signals the PLC through input X0 of this condition. An accumulating timer is activated through X0 accumulating the time X0 is on. X0 can turn on and off, but each time it turns on the accumulating timer keeps track. After 10 seconds of accumulation, the timer turns on Y0, which turns on an alarm. X1 is the reset for the timer. Refer to Figure 4.17.

To recap:

X0—Start timer

X1—Reset timer

Y0—Output

Equipment Required

1. PLC
2. 12V power supply
3. Buzzer
4. Push button switches

To demonstrate the lab, apply 12VDC to X0 and observe the timer count; remove the 12V from X0 and apply again. Notice how the time is retained. Observe the output Y0 if X0 is activated long enough.

Questions

1. If the timer's count is zero and X0 is activated for 10 seconds, what does Y0 do?
2. If the timer's count is 8s and X0 is activated for 2 seconds, what does Y0 do?
3. If the timer's count is 10s and X0 is activated for 2 more seconds, what does Y0 do?
4. If the sensor is working, sending a signal to X0 for 10s, and the alarm never turns on, what could be the problem with X1?
5. If the sensor is working, sending a signal to X0 for 10s, and X1 is fine and the alarm never turns on, what could be the problem?

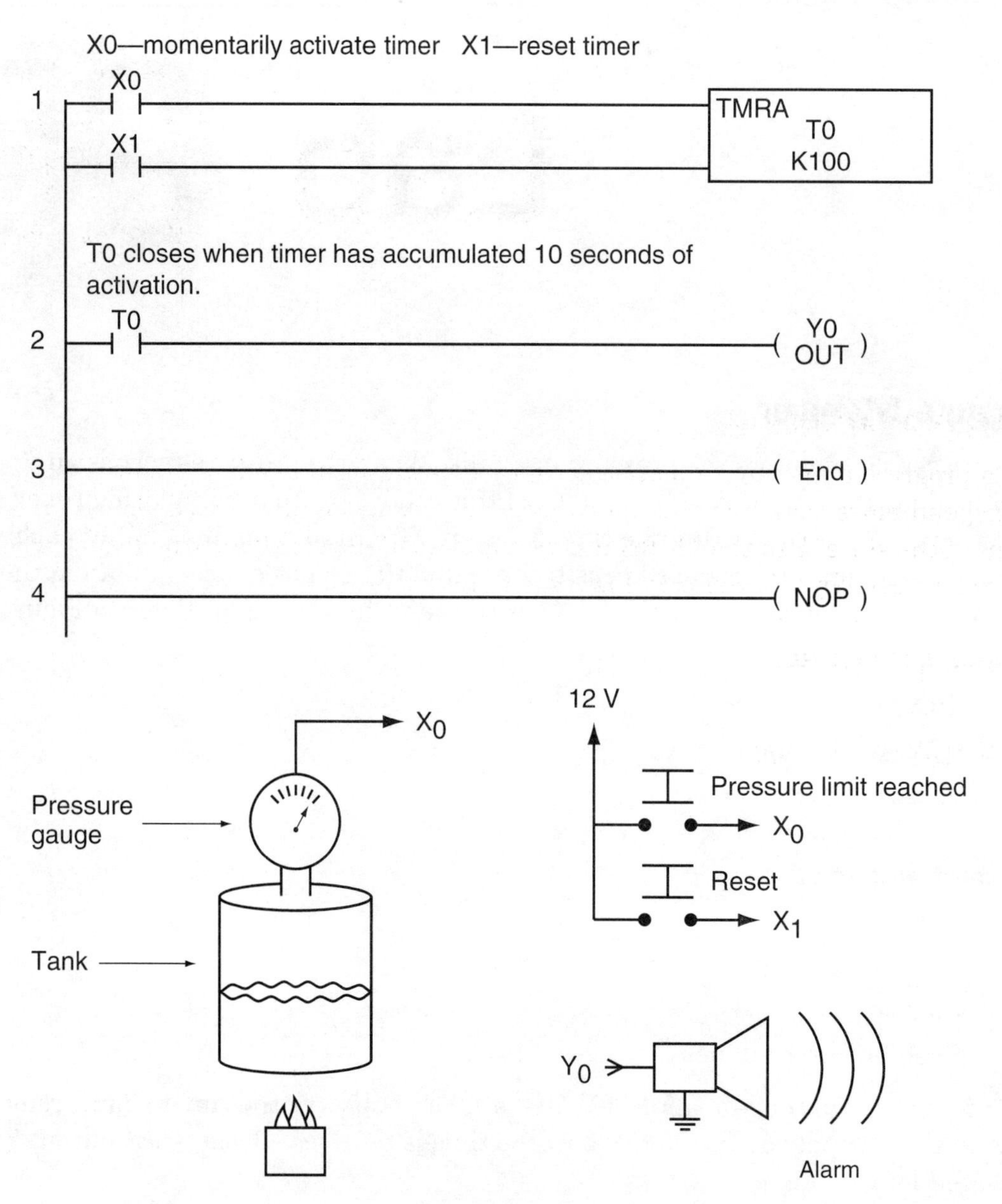
05 Pressure monitor
X0—momentarily activate timer X1—reset timer
X0
1
TMRA
T0
K100
X1
T0 closes when timer has accumulated 10 seconds of activation.
T0
2
Y0
OUT
3
End
4
NOP
12 V
X_0
Pressure gauge
Pressure limit reached
X_0
Reset
X_1
Tank
Y_0
Alarm

Figure 4.17

Name ______________________________ Date ______________

Lab 

Write No Loitering

Write a program to monitor a motion sensor located in front of a liquor store. When the loiterers spend more than 12s in front of the store, have classical music played on speakers in front of the store. Use the TMRA function and have the speaker turned on by Y0 and the sensor monitored by X0. Refer to Figure 4.17 for help.

Equipment Required

1. PLC
2. 12V power supply
3. Buzzer
4. Push button switch

Name ______________________________ Date ______________

Lab

Temperature Controlled Fan

A bimetal strip temperature switch monitors the temperature of an environment. It is designed so that when it warms it bends downward, closing the switch and delivering 12V to X0. TMRA is the timer accumulate function that will turn on T0 after X0 has been on an accumulated time of 5 seconds. Timer T0 then turns on timer T2 for 10 seconds, which turns on Y2, the fan. When T2 reaches 10 seconds, it will reset timer T0 which will then reset timer T2. Refer to Figure 4.18.

To recap:

X0—Bimetal temperature switch

Y2—Fan power

Equipment Required

1. PLC
2. 12V power supply
3. 12VDC motor
4. Bimetal temperature switch or toggle switch to simulate

To demonstrate the lab, apply12VDC to X0 and observe the timer count; remove the 12V from X0 and apply again. Notice how the time is retained. Observe the output if X0 is activated long enough.

Questions

1. What happens if X0 is activated for 5 seconds?
2. If the fan is on, what timer turns it off?
3. What can be changed in the program to keep the fan on longer?
4. What can be changed in the program to turn the fan on sooner?
5. What could cause the fan never to turn on?
6. What is the difference in behavior between TMR and TMRA?

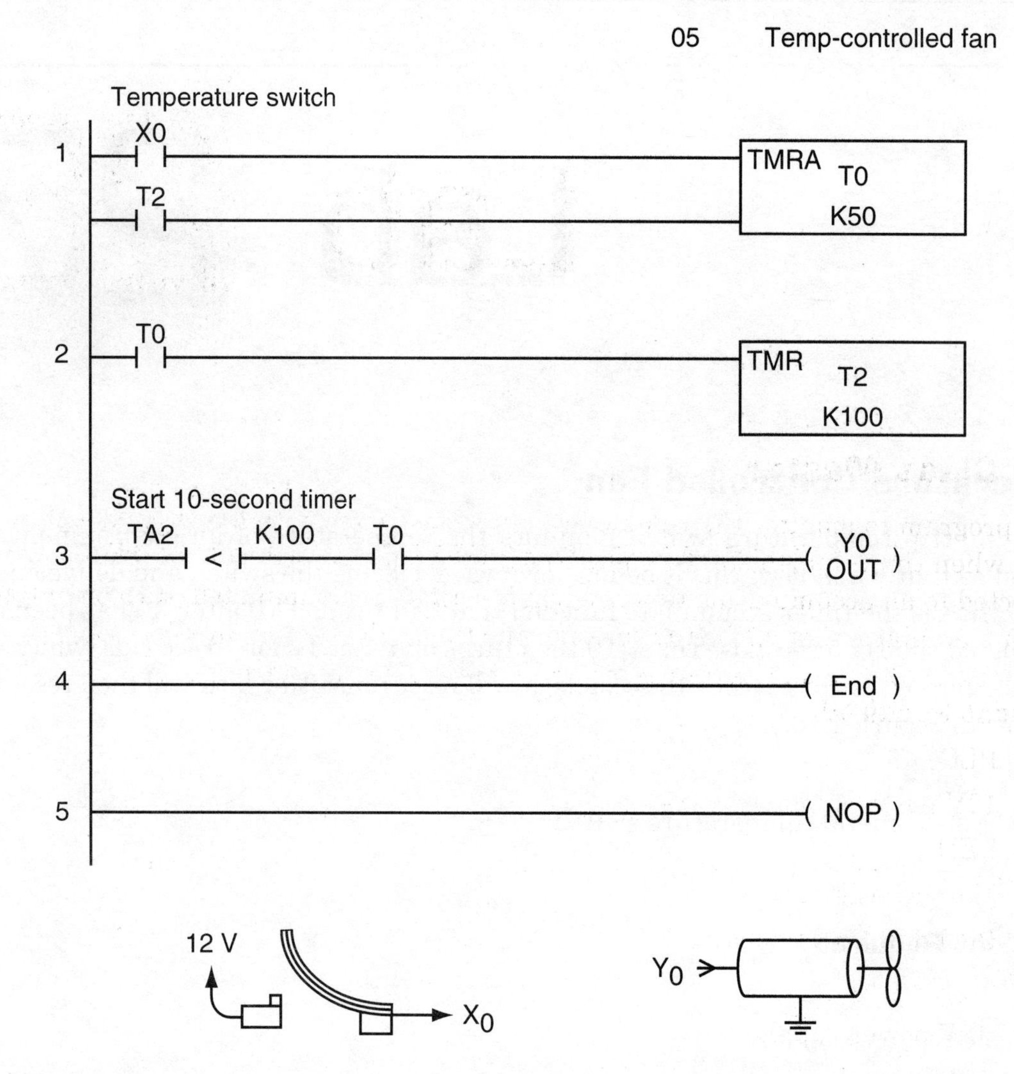
05 Temp-controlled fan
Temperature switch
X0
T2
TMRA T0 K50
T0
TMR T2 K100
Start 10-second timer
TA2
<
K100
T0
Y0 OUT
End
NOP
1
2
3
4
5
12 V
X0
Y0

Figure 4.18

Name ______________________________ Date ______________

Lab 

Write Class Master

Write a program to monitor the volume level in a class room with a microphone that puts out 12V when the volume gets too high. This microphone is connected to input X0, which is connected to an accumulating timer. When 5 seconds is accumulated, turn on output Y0 to activate an alarm. Let X1 be the reset for the timer. Refer to Figure 4.17 for help.

Equipment Required

1. PLC
2. 12V power supply
3. LED
4. 1kohm resistor
5. Push button switches

SECTION 5 Drum Sequencers

Objective

Upon completion of these labs, you should be able to:

- Understand how a drum sequencer can be used to control a system.

Introduction

A drum sequencer simulates an electro-mechanical drum sequencer.

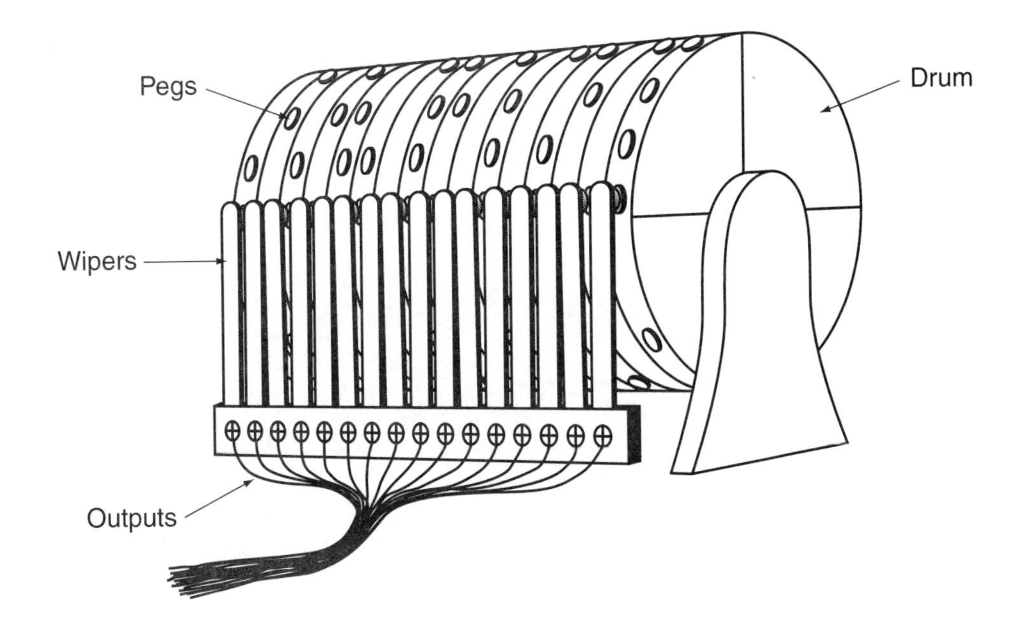

Figure 5.1 An electro-mechanical drum sequencer.

Drum sequencers are in many cases much easier to program with and more efficient in terms of fewer program rungs than the TMR function. The drum sequencer function outputs such as Y0, Y1, etc., are labeled in the function.

Look at Figure 5.2.

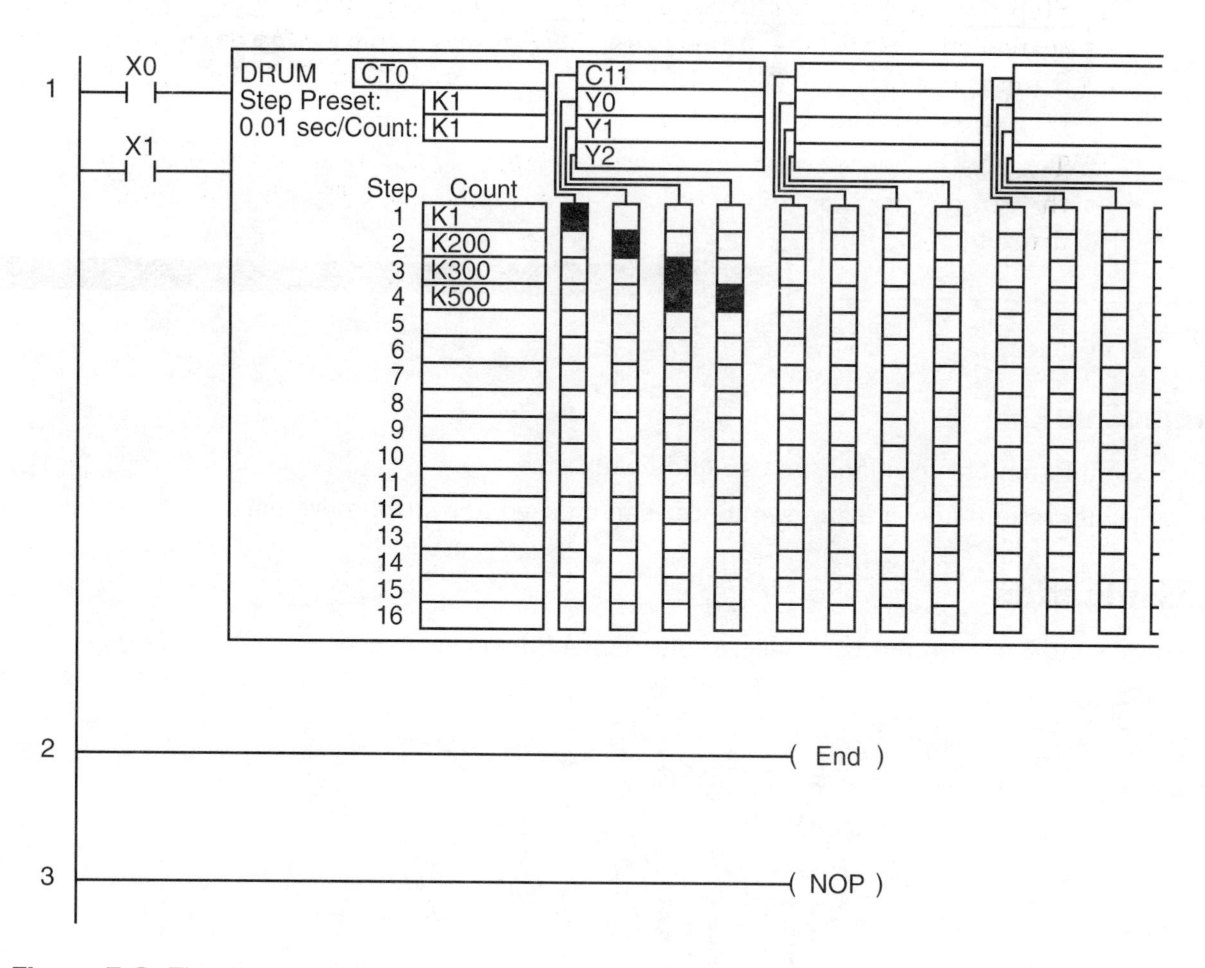

Figure 5.2 The drum sequencer.

In Figure 5.2, the drum's outputs are labeled in the upper right boxes of the drum by C11, Y0,Y1, and Y2. The drum is started by activating the start input X0. When the drum is activated, the step preset is referenced and tells the drum which step to begin on. The time the drum spends on each step is determined by the box 0.01 sec/count and the number in the Count box. The box 0.01 sec/count with k1 means each count will last 0.01s. If the count box contains k200, then the time spent on step 2 is k1*0.01sec/count*k200 = 2 seconds. The time spent on step 3 is k1*0.01sec/count*k300 = 3 seconds. During each step the output connected to that step will be activated.

STEP 1. C11 will be activated for 0.01s.

STEP 2. Y0 will be on for 2s.

STEP 3. Y1 will be on for 3s.

STEP 4. Y1 and Y2 will be both on for 5s.

When all the steps have finished, the drum's output will turn on, activating any switch labeled CT0. The drum can be reset at any time with the reset input controlled by X1.

Drum Subtleties

1. Step one, by default, on the drum, will always be activated before the drum is turned on. This means in this example, Figure 5.2, that C11 will be on before the drum is instructed to begin. Using a dummy variable like C11 prevents one of the Y outputs from being turned on.
2. To have the drum cycle back on after completing all the steps, have the output of the drum CT0 connected to the reset. On completion, CT0 will close, resetting the drum to begin over.

Name ______________________________ Date ______________

Lab 29

Washing Machine Drum

A washing machine runs through the following cycle: FILL, AGITATE, DRAIN, and then SPIN. The program begins when the start button X0 is pressed. The machine begins to fill by opening the fill valve for 2 seconds, then agitates for 2 seconds, then drains for 2 seconds, and finally spins for 2s with the drain remaining open while spinning. Refer to Figure 5.3.

To recap:

X0—Start
X1—Stop
Y0—Fill valve
Y1—Agitate
Y2—Drain valve
Y3—Spin

Equipment Required

1. PLC
2. 12V power supply
3. LEDs
4. 1kohm resistors
5. Push button switches

To demonstrate the lab, momentarily apply 12VDC to X0 to start, and X1 to reset, and observe the outputs.

Questions

1. What happens to C0 when X0 is pressed?
2. What happens during the 2s to 4s period of the cycle?
3. If the machine never fills and the program is working fine, what could be the problem?
4. If the machine never drains and the program is working fine, what could be the problem?

5. If the machine never spins and the program is working fine, what could be the problem?
6. What turns off the timer in the ladder logic diagram?
7. How long is the drain open for?
8. How would you distinguish between a problem with the machine's motor and a problem with the PLC?

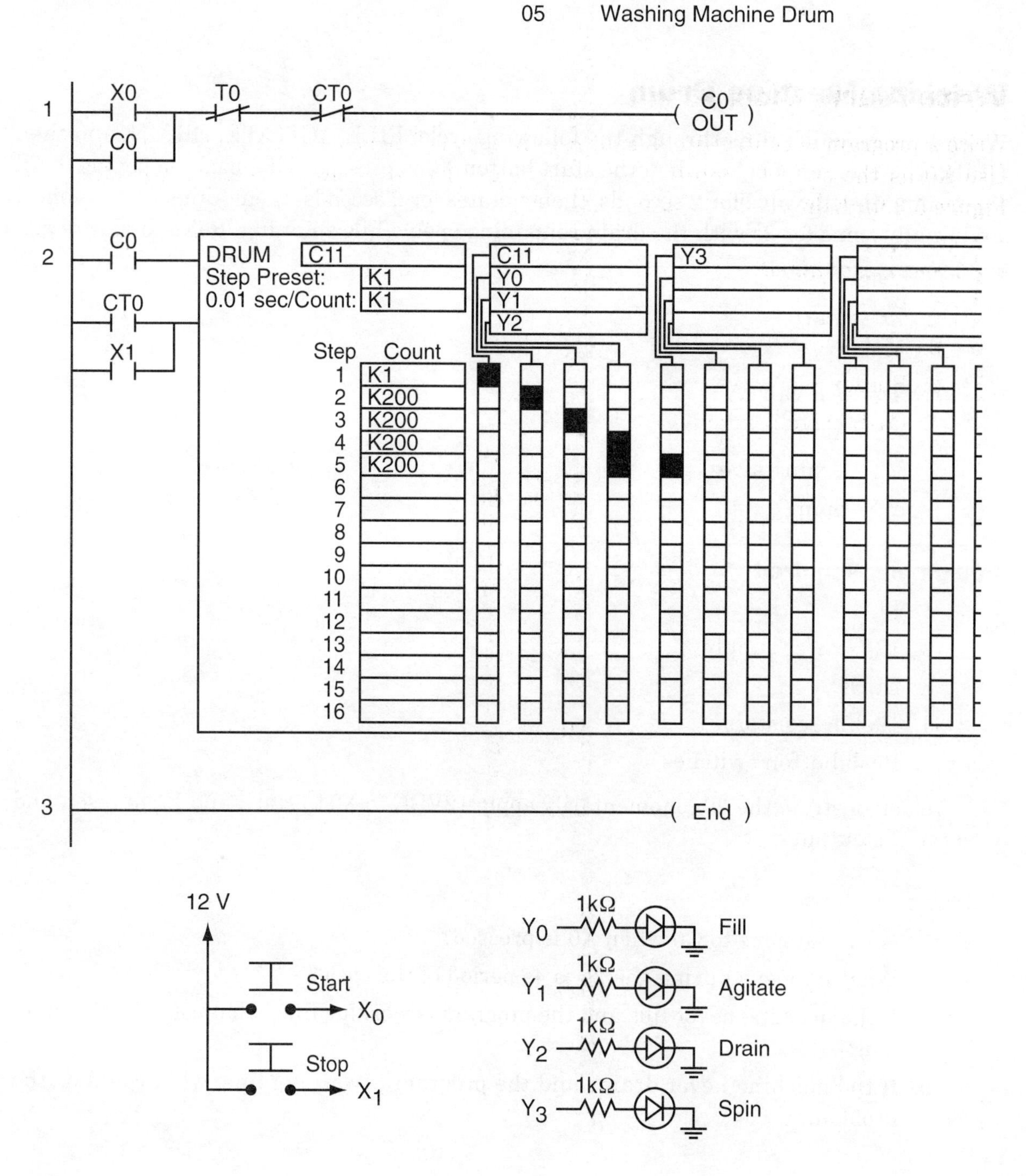

Figure 5.3

Name ______________________ Date ______________

Lab 

Write Alarm Clock Drum

Write a program to control an alarm clock that rings 8 seconds after it has been activated. Use X0 as the start button, X1 as reset, and Y0 as the output to the ringer. Refer to Figure 5.3 for help.

Equipment Required

1. PLC
2. 12V power supply
3. Buzzer
4. Push button switches

Name ______________________ Date ______________

Lab 

Drum Traffic Light

The green light on the traffic light is controlled by output Y0 and is on for 3 seconds, the yellow light by Y1 and is on for 3s, and the red light by Y2 and is on for 4s. The timer is self-starting and self-resetting by CT0. Refer to Figure 5.4.

To recap:

Y0—Green light

Y1—Yellow light

Y2—Red light

Equipment Required

1. PLC
2. 12V power supply
3. LEDs
4. 1kohm resistors

To demonstrate the lab, run the program and observe the time sequence of the outputs.

Questions

1. Draw a timing diagram showing when the green, yellow, and red lights are on.
2. If the lights never come on, what could the problem be?
3. What causes the timer to reset?

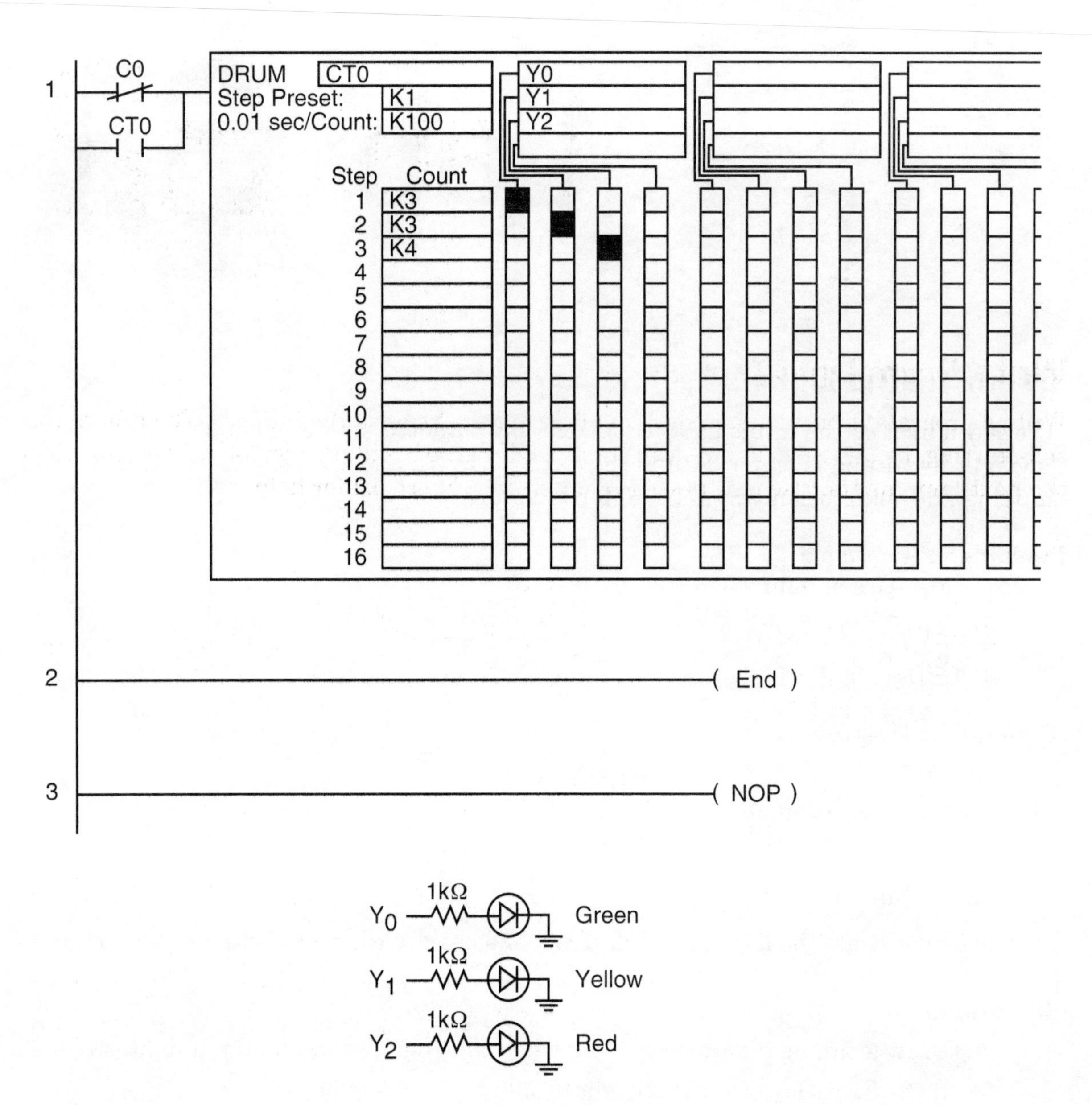
05 Drum Traffic Light
1
C0
CT0
DRUM CT0
Step Preset: K1
0.01 sec/Count: K100
Y0
Y1
Y2
Step Count
1 K3
2 K3
3 K4
4
5
6
7
8
9
10
11
12
13
14
15
16
2
(End)
3
(NOP)
1kΩ
Y0
Green
1kΩ
Y1
Yellow
1kΩ
Y2
Red

Figure 5.4

Name ______________________________ Date ______________

Lab 

Write Drum Light Chaser

Write a program to turn on and off 6 lights in a sequence and then repeat. The program is self-starting, i.e., it needs no input to start. Each light should turn on for 0.5 seconds, then the next lamp lights. Use outputs Y0–Y5. Refer to Figure 5.4 for help.

Equipment Required

1. PLC
2. 12V power supply
3. LEDs
4. 1kohm resistors

Name ______________________ Date ______________

Lab 

Drum Traffic + Walk Light

The green light on the traffic light is controlled by output Y1 and is on for 3 seconds, the yellow light by Y2 and is on for 3s, and the red light by Y3 and is on for 3s. The timer is self-starting and self-resetting by CT0. A flashing walk light connected to Y0 flashes at a rate of one flash every half-second while the green light is on. Refer to Figure 5.5.

To recap:

Y0—Flashing walk light

Y1—Green light

Y2—Yellow light

Y3—Red light

Equipment Required

1. PLC
2. 12V power supply
3. LEDs
4. 1kohm resistors

To demonstrate the lab, run the program and observe the time sequence of the outputs.

Questions

1. Draw a timing diagram showing when the green, flashing, yellow, and red lights are on.
2. If the lights never come on, what could the problem be?
3. What causes the timer to reset?

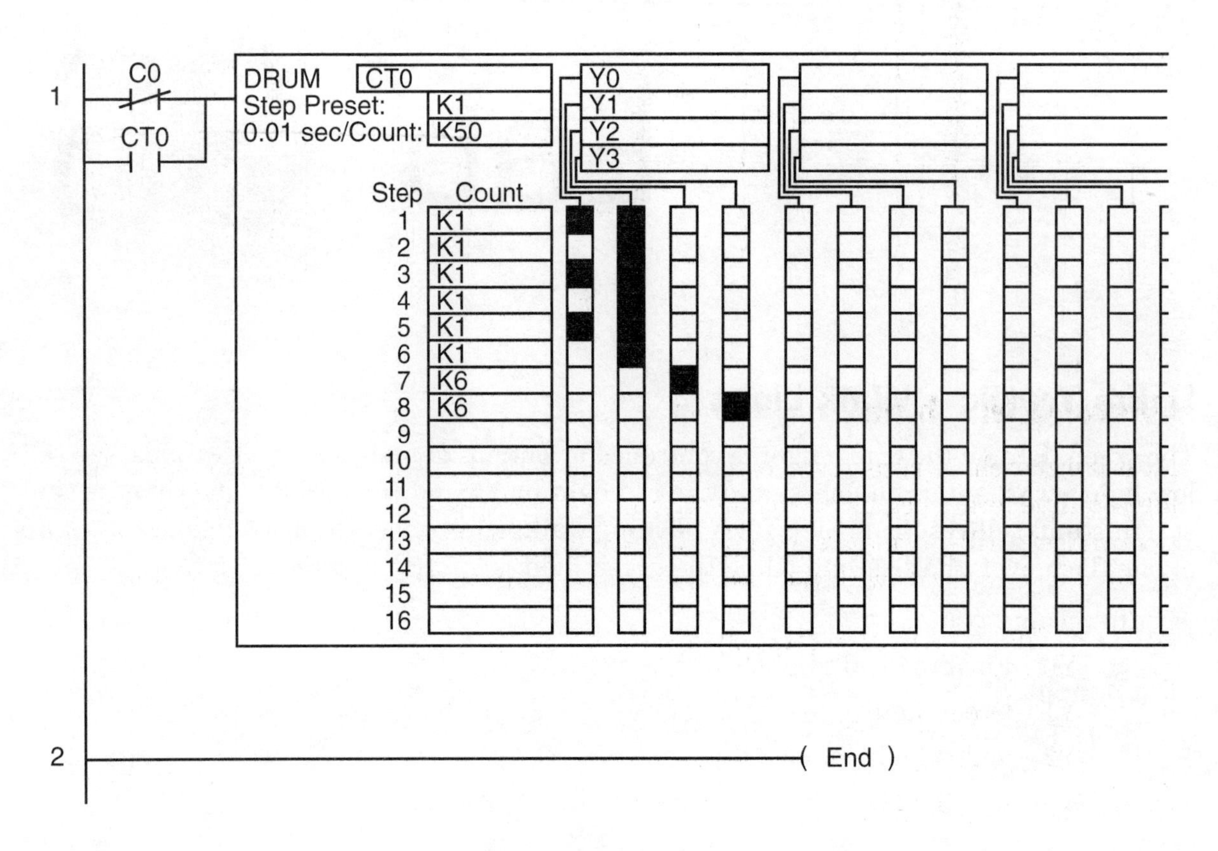
05 Drum Traffic & Walklight
1
C0
CT0
DRUM CT0
Step Preset: K1
0.01 sec/Count: K50
Y0
Y1
Y2
Y3
Step Count
1 K1
2 K1
3 K1
4 K1
5 K1
6 K1
7 K6
8 K6
9
10
11
12
13
14
15
16
2
End

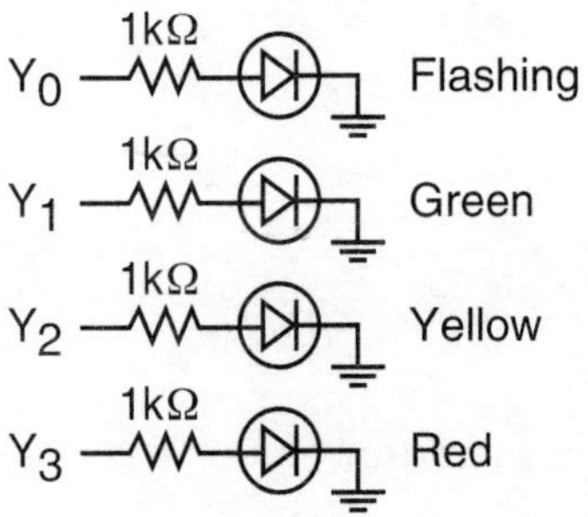
1kΩ
Y0
Flashing
1kΩ
Y1
Green
1kΩ
Y2
Yellow
1kΩ
Y3
Red

Figure 5.5

Name ______________________ Date ______________

Lab

Write Drum Burglar Alarm

Most burglar alarms have a delay on. Once you press the on button, a delay takes place giving you time to close the door before arming the alarm. Write a program using the drum function to arm an alarm 5 seconds after pressing the X0 start button. X1 is the reset, and Y0 is the output to the siren. Let X2 trigger the alarm after it is armed. Once the alarm has been triggered by X2, have it stay on. Refer to Figure 5.4 (page 84) for help.

Equipment Required

1. PLC
2. 12V power supply
3. Buzzer
4. Push button switches

SECTION 6

Counters

Objectives

Upon completion of these labs, you should be able to:

- Understand how PLCs use up and up/down counters to control a system.

Introduction

Counters keep track of how many times an input has been activated. They are used in applications such as counting the number of objects passed along a conveyor belt or placed into a container. There are two counter functions we will discuss, count abbreviated CNT, Figure 6.1, and up down counter, abbreviated UDC. The CNT function is a counter that has two inputs, count up and reset, and an output that turns on when the count is reached. A number such as k5 means 5 counts.

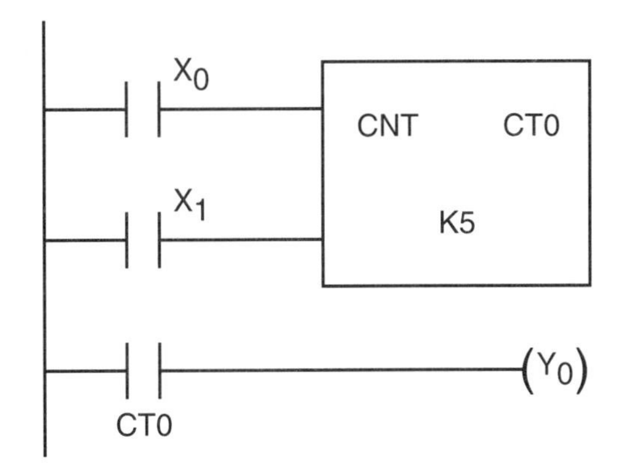

Figure 6.1 Activating X0 5 times will turn on CT0, closing contact CT0, and turn on Y0. Activating X1 will reset the timer.

The up down counter, UDC, has three inputs, up, down, and reset, Figure 6.2. The up count will increment the count, the down count will decrement the count, and reset will zero out the count.

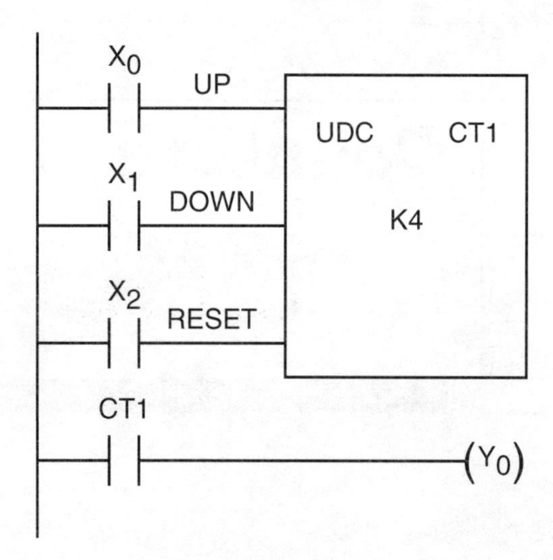

Figure 6.2 The counter will turn on the output CT1 and hence the output Y0 when the count reaches 4, k4. X0 is connected to the up count, and X1 to the down count. Activating X0 will increment the count. Activating X1 will decrement the count. Activating X2 will reset the count to zero.

Name ______________________________ Date ______________

Lab

Rain Drop Counter

A rain drop counter on a windshield counts rain drops with a sensor connected to X0, which is connected to a counter. The counter has a reset X1. When the count reaches 20, contact CT0 closes and output Y0 turns on, activating a wiper. Refer to Figure 6.3.

To recap:

X0—Count up

X1—Reset

Y0—Output

Equipment Required

1. PLC
2. 12V power supply
3. Push button switch
4. Motor

To demonstrate the lab, momentarily apply 12V to X0 multiple times, and observe the count. Allow the count to reach the terminal count and observe the output.

Questions

1. If the internal count of the counter is 18 and 2 more raindrops are recorded, what happens?
2. If X0 is activated, what happens to the count?
3. If X1 is activated, what happens to the count?
4. If the counter never counts, what could be happening with X0?
5. If the wiper never turns on and the count is 25, what could be the problem?

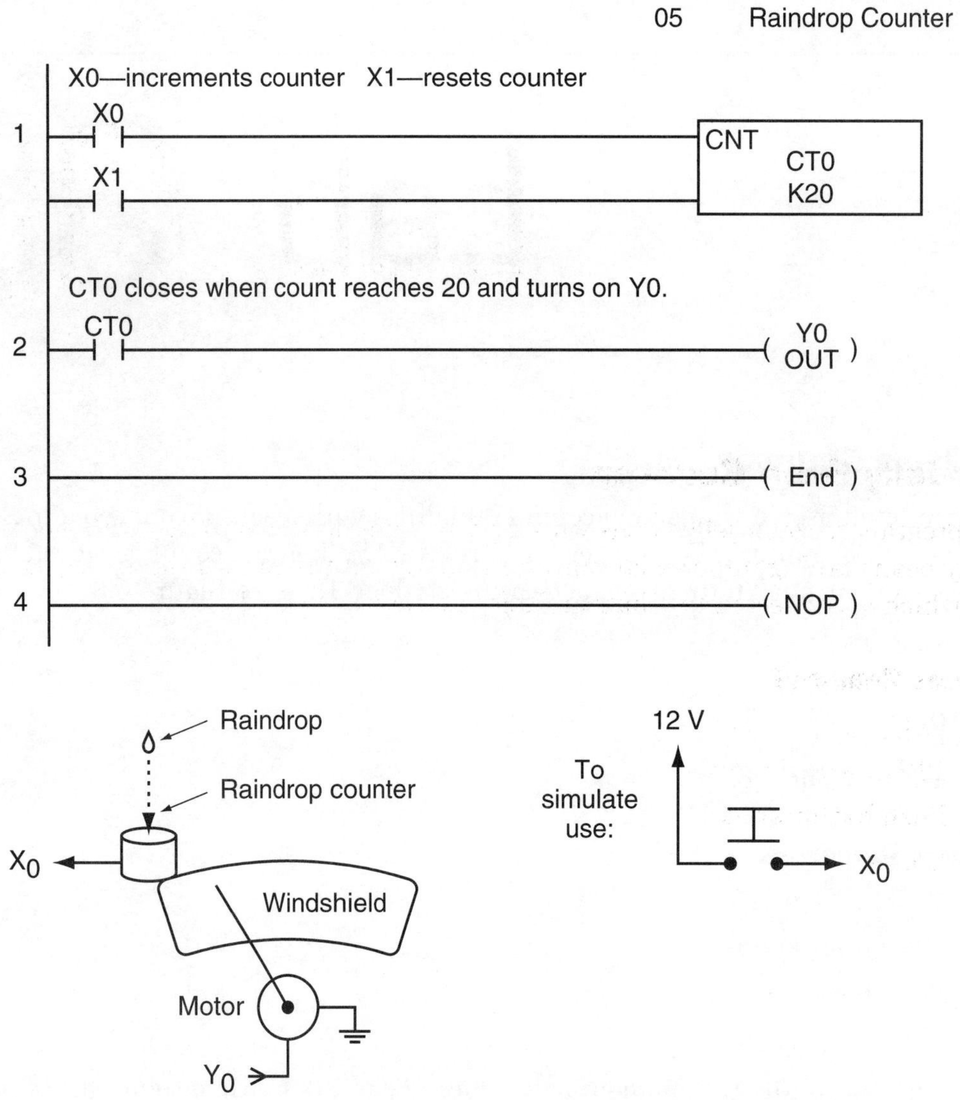

05 Raindrop Counter

X0—increments counter X1—resets counter

1 X0 —| |— CNT CT0 K20
X1 —| |—

CT0 closes when count reaches 20 and turns on Y0.

2 CT0 —| |— (Y0 OUT)

3 (End)

4 (NOP)

Raindrop
Raindrop counter
X_0
Windshield
Motor
Y_0

To simulate use:
12 V
X_0

Figure 6.3

Name ______________________________ Date ______________

Lab 

Write Jelly Bean Counter

Write a program to count jelly beans with a photo gate connected to X0 positioned over a jar with jelly beans falling into the jar one at a time. When the count reaches 18, turn on output Y0, which is connected to a buzzer. Refer to Figure 6.3 for help.

Equipment Required

1. PLC
2. 12V power supply
3. Push button switch
4. Buzzer

Name ______________________________ Date ______________

Room Capacity Alarm

A room capacity alarm monitors the number of people in a room and sets off an alarm when the room capacity is exceeded. This could be done with a set of photo gates at the doorway. Input X0 counts the number of people entering and X1 the number of people leaving. The up/down counter UDC keeps track of the total number in the room. When a count of 10 is reached, the counter's output CT0 turns on, closing switch CT0, and turns on output Y0 activating the alarm. Input X2 resets the counter. Refer to Figure 6.4.

To recap:

X0—Count up

X1—Count down

X2—Reset

Y0—Output

Equipment Required

1. PLC
2. 12V power supply
3. Push button switches
4. Buzzer

To demonstrate the lab, momentarily apply 12V to X0 and X1 multiple times, and observe the count. Allow the count to reach the terminal count and observe the output.

Questions

1. If the room is empty and 5 people enter the room, then 2 people leave the room, what is the internal count of the counter?
2. What happens if a count of 10 is reached?
3. Can the alarm be turned off using X2?
4. Can the alarm be turned off using X1?
5. If the alarm never turns on even with 15 people in the room, what could be the problem with X0?
6. If the alarm is on then some people leave, bringing the room count below ten, the alarm should go off. If this does not happen, what could be wrong?

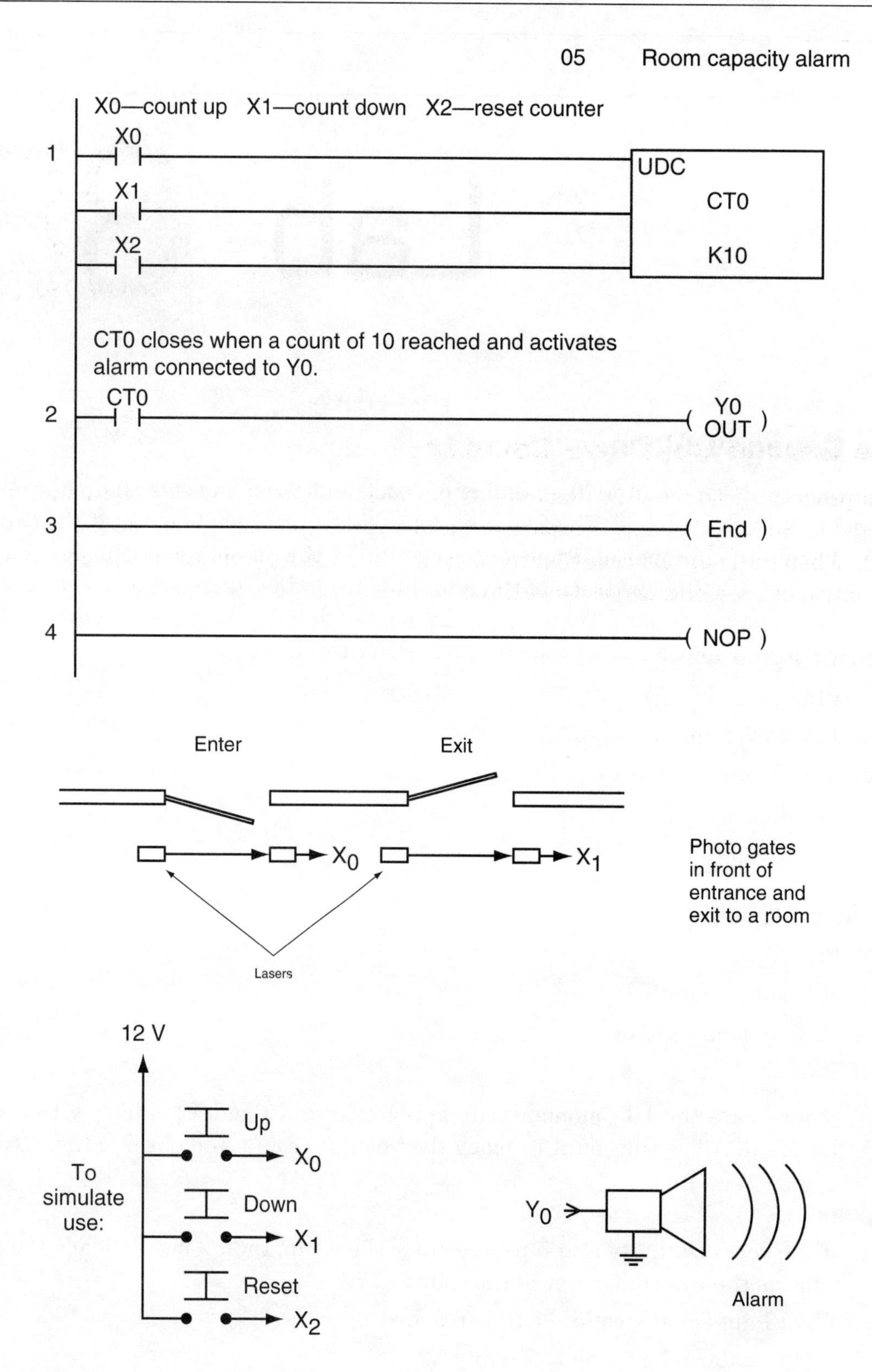
05 Room capacity alarm
X0—count up X1—count down X2—reset counter
1
X0
X1
X2
UDC
CT0
K10
CT0 closes when a count of 10 reached and activates
alarm connected to Y0.
2
CT0
Y0
OUT
3
End
4
NOP
Enter
Exit
X0
X1
Photo gates
in front of
entrance and
exit to a room
Lasers
12 V
To
simulate
use:
Up
X0
Down
X1
Reset
X2
Y0
Alarm

Figure 6.4

Name ______________________ Date ______________

Lab

Write Garage Up/Down Counter

Write a program to count the number of cars entering a parking garage with a photo gate connected to X0 positioned at the entrance, and a photo gate connected to X1 positioned at the exit. When the count reaches 8, turn on output Y0, which is connected to a gate that closes the entrance gate and leaves the exit open. Refer to Figure 6.4 for help.

Equipment Required

1. PLC
2. 12V power supply
3. Push button switches
4. Motor

SECTION 7

Motor Control

Objective

Upon completion of these labs, you should be able to:

- Understand how a PLC can be used to control both the speed and direction of rotation of a motor.

Introduction

A motor's speed and rotation direction will be controlled in the following labs. We will introduce new commands in these labs to achieve this control.

The "SET" and Reset "RST" Command (LAB 39, 42)

The SET command turns on an output and keeps it on until it is reset by the RST command, Figure 7.1.

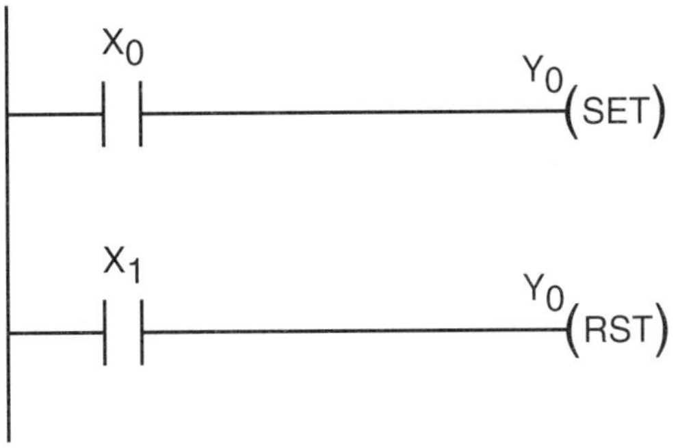

X_0 — Y_0 (SET)

X_1 — Y_0 (RST)

Figure 7.1 Activating X0 turns on Y0. Y0 will remain on even if X0 is deactivated because Y0 has been "SET." To turn off Y0 it must be reset using RST. Y0 is reset if X1 is activated, turning off Y0.

Double Pole Double Throw Relay (DPDT Relay)

The 12V DPDT relay used in these labs has a pin-out shown in the Figure 7.2. See the spec sheet in Appendix B, Figure B.5, for this relay. Use an ohm meter to determine the pin out for your relay.

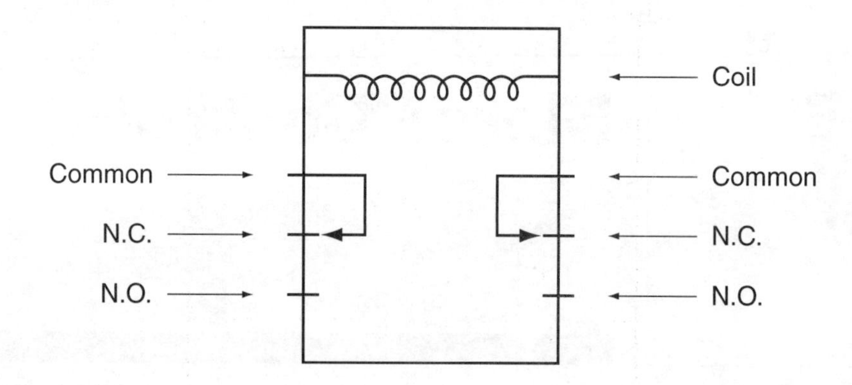

Figure 7.2 X-Ray view of a DPDT relay.

Inductive Loads

Motors and relay coils are inductive loads; that is, they are made of coils of wire. When an inductive load with current running through it has the current removed, a large voltage spike is developed across it. This is called inductive kickback, and could damage circuits connected to it. To protect against this, a diode can be placed across the load in a reversed bias manner as shown in Figure 7.3.

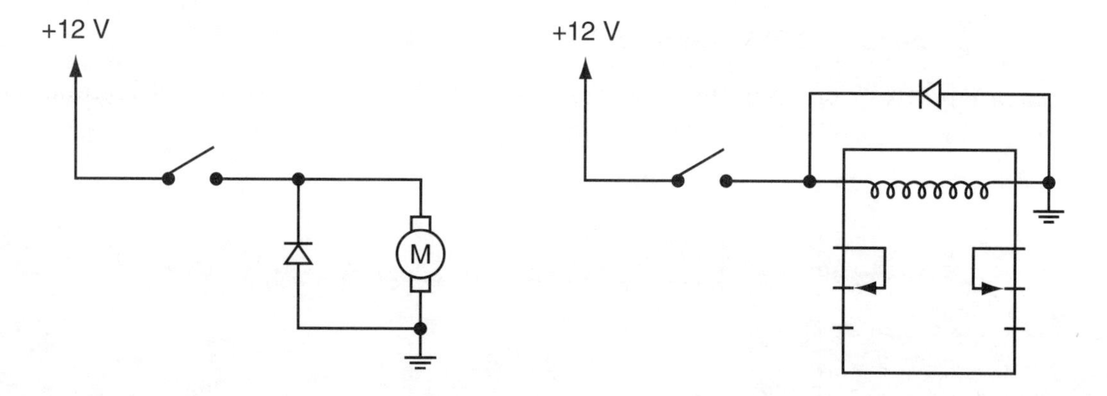

Figure 7.3 A diode placed across an inductive load such as a motor or a relay coil will help protect any circuit connected to it from inductive kickback. Notice the anode is facing toward ground.

Controlling a motor's direction of rotation is typically done with a circuit called an H-Bridge. By opening and closing the contacts in the right order, the polarity of the voltage on the motor can be reversed hence reversing the motor's rotation. See Figure 7.4. To have the motor spin in one direction, close contacts A and D; to have it reverse, close contacts C and B. Never should contacts A and B or C and D be closed at the same time. This would, of course, short out the power supply.

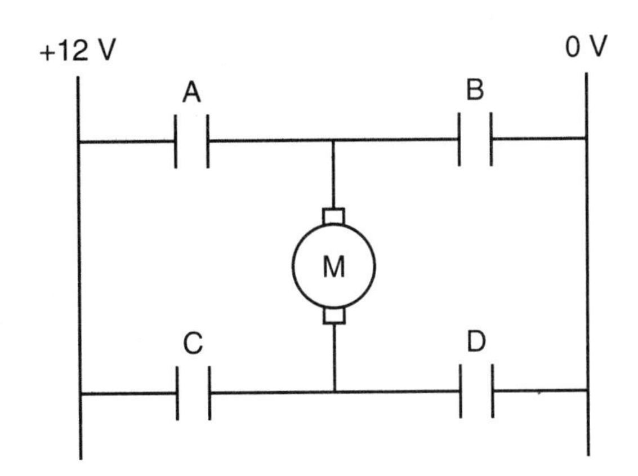

Figure 7.4 An H-Bridge. To have the motor spin in one direction close contacts A and D; to have it reverse, close contacts C and B. Never should contacts A and B or C and D be closed at the same time.

Name ______________________ Date ______________

Lab

Toggle Motor On/Off

A single momentary switch X0 controls a motor, either on or off. A counter counts the number of times X0 is activated, and resets itself after 2 counts. Y2 is either set (turned on) or reset (turned off) when the count is one or two respectively. Refer to Figure 7.5.

To recap:

X0—Turn on or off motor

Y2—Motor power

Equipment Required

1. PLC
2. 12V power supply
3. 12V motor
4. Push button switch

To demonstrate the lab, momentarily apply 12VDC to X0, while observing the output, then repeat.

Questions

1. If the motor is off and X0 is activated, what happens?
2. If the motor is on and X0 is activated, what happens?
3. Does X0 have to remain pushed for the motor to stay on or off?
4. If the motor gets stuck on or off, what could be the problem? Assume PLC OK and outputs are working.
5. What does the SET command do in the program?
6. What does the RST command do in the program?
7. For what count does Y2 turn on?
8. For what count does Y2 turn off?

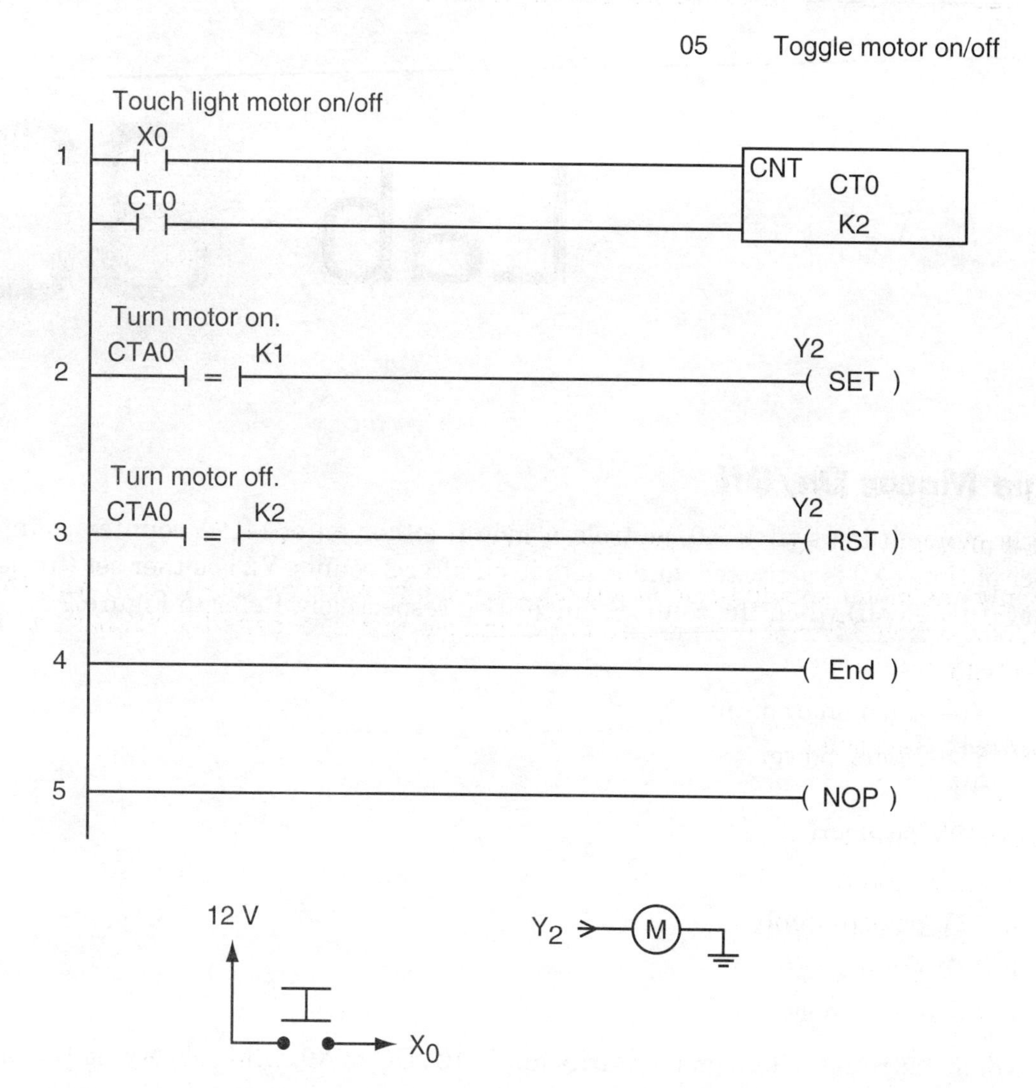
05 Toggle motor on/off
Touch light motor on/off
1
X0
CT0
CNT
CT0
K2
Turn motor on.
2
CTA0
K1
Y2
SET
Turn motor off.
3
CTA0
K2
Y2
RST
4
End
5
NOP
12 V
X0
Y2
M

Figure 7.5

Name ________________________________ Date ______________

Lab 

Write Toggle Motors

Write a program to toggle two motors on and off using one switch X0.

Only one motor should be on at a time. Each motor will alternately be on with the activation of a switch. Motor 1 is connected to Y1, and motor 2 is connected to Y2. Refer to Figure 7.5 for help.

Equipment Required

1. PLC
2. 12V power supply
3. 12V motors
4. Push button switches

Name ______________________________ Date ______________

Lab

Forward/Reverse Motor Control

This motor control program is designed to change the direction of rotation of a motor using an H-Bridge. X1 selects the forward direction and X2 the reverse. These two switches are latched so that they can be selected with a momentary switch. X0 is the stop switch. The program is designed so that forward and reverse cannot be selected at the same time. One of the motor's leads should be connected to the common of outputs Y0 and Y1, called C2, the other lead should be connected to the common of outputs Y3 and Y4, called C3. Y0 and Y4 should be connected to 12V, and Y1 and Y3 to ground. Refer to Figure 7.6.

To recap:

X0—Stop
X1—Forward
X2—Reverse
Y0—12V
Y1—0V
Y3—0V
Y4—12V
C2—One of the motor's leads
C3—The other motor lead

Equipment Required

1. PLC
2. 12V power supply
3. 12V motor
4. Push button switches

To demonstrate the lab, apply 12VDC to forward X1 while observing the rotation of the motor. Next, stop the motor by applying 12V to X0. Reverse the motor by applying 12V to X2.

Questions

1. What happens if X0 is selected?

2. What happens if X1 is selected?
3. What happens if X2 is selected?
4. What happens if the motor is running in the forward direction and the reverse button is pressed?
5. If the motor is running in the forward direction, what switch or switches need to be activated to get it to reverse?
6. What could be the problem if the motor will not go into reverse?

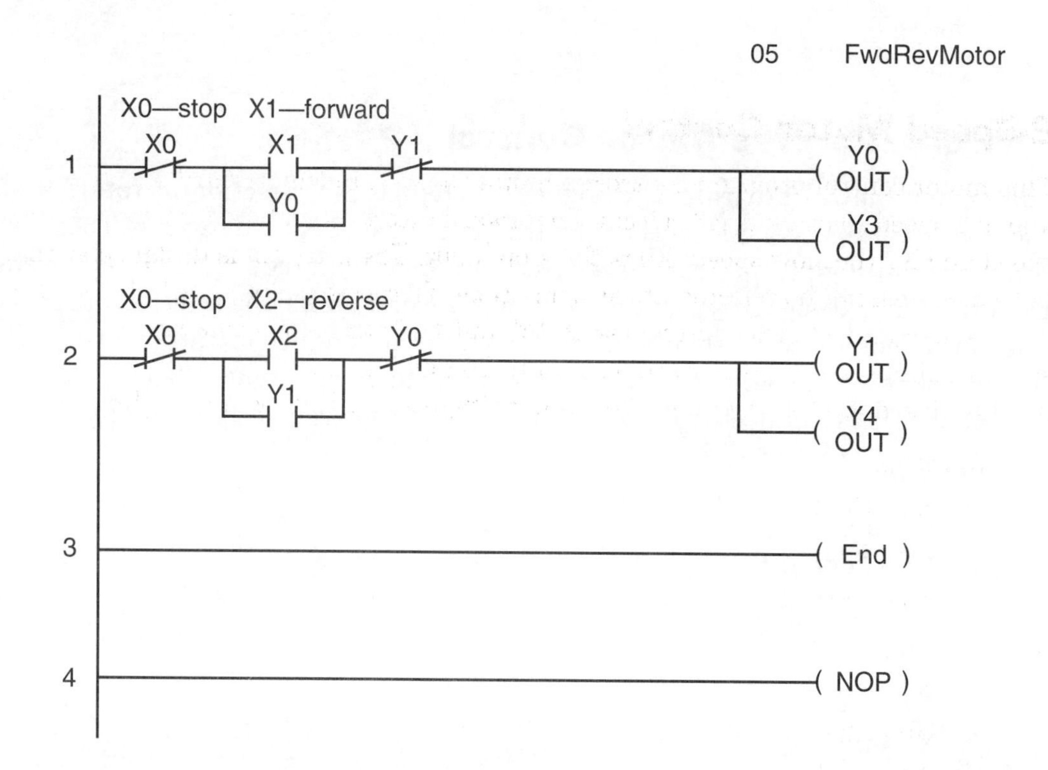

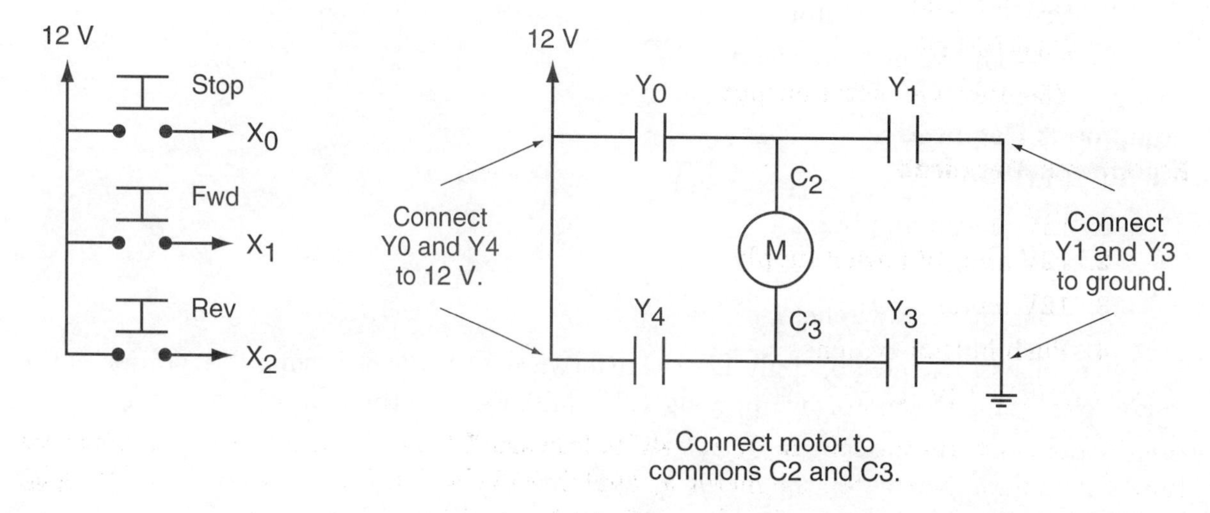

Figure 7.6

Name ______________________ Date ____________

Lab 42

2-Speed Motor Control

This motor control program controls a motor to move in the forward and reverse directions and is 2 speed selectable. X3 selects the forward direction and X4 reverse. X1 selects the fast speed and X2 the slow speed. X0 is the stop input. The program is designed so that slow and fast speeds cannot be selected at the same time. Y3 controls a relay, which changes the polarity on the motor in order to reverse it. Y0 and Y1 both have a common C2. To obtain the 2 speeds, different voltages are connected to Y0 and Y1. C2 will have the same voltage as Y0 if Y0 is closed, or C2 will have the same voltage as Y1 if Y1 is closed. Refer to Figure 7.7.

To recap:

X0—Stop
X1—Fast
X2—Slow
X3—Forward
X4—Reverse
Y0—12V in
Y1—6V in
C2—To relay
C3—To 12V
Y3—Reverse motor output

Equipment Required

1. PLC
2. 12V and 6V power supply
3. 12V motor
4. Push button switches
5. DPDT 12V relay

To demonstrate the lab, apply 12VDC to forward X3 and X1 while observing the rotation of the motor. Next, stop the motor by applying 12V to X0. Reverse the motor by applying 12V to X4. Choose fast or slow by applying 12V to X1 or X2 respectively.

Questions

1. If X1 and X3 are activated, what does the motor do?
2. If X2 and X3 are activated, what does the motor do?
3. If X1 and X4 are activated, what does the motor do?
4. If X2 and X4 are activated, what does the motor do?
5. If the motor is running and X0 is activated, what does the motor do?
6. If the motor never turns on, what could be the problem?
7. If the motor never reverses, what could be the problem?
8. If the motor only reverses, what could be the problem?
9. If the motor never turns off, what could be the problem?

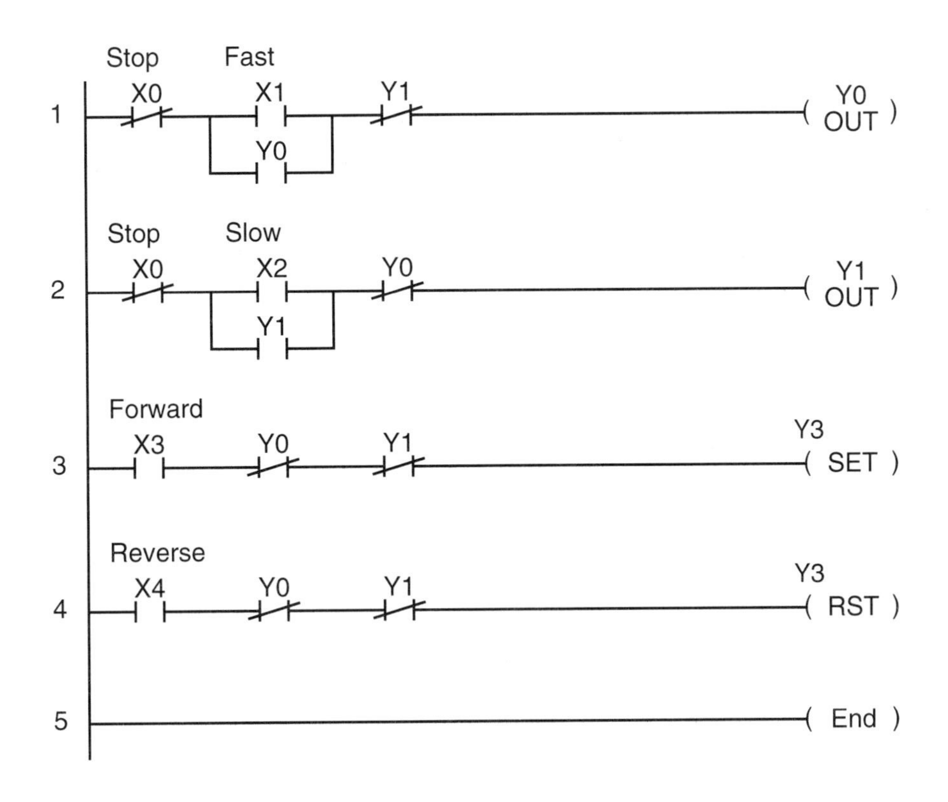
05 2-Speed motor
Stop
Fast
X0
X1
Y1
Y0
OUT
Y0
Stop
Slow
X0
X2
Y0
Y1
OUT
Y1
Forward
X3
Y0
Y1
Y3
SET
Reverse
X4
Y0
Y1
Y3
RST
End

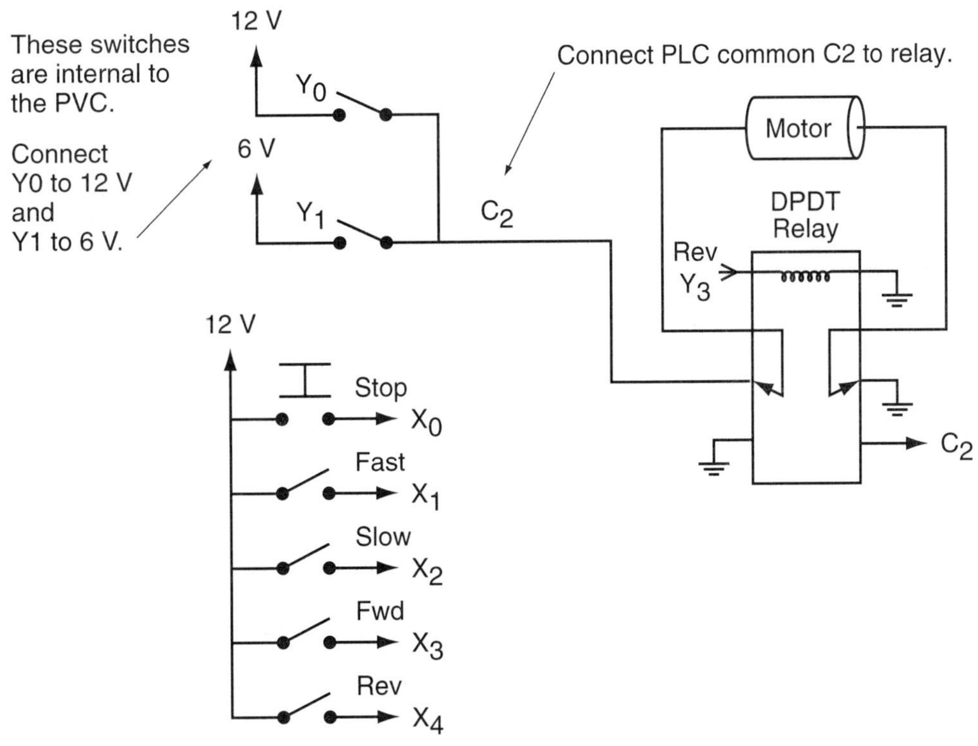
These switches are internal to the PVC.
Connect Y0 to 12 V and Y1 to 6 V.
12 V
6 V
Y0
Y1
C2
Connect PLC common C2 to relay.
Motor
DPDT Relay
Rev
Y3
C2
12 V
Stop
X0
Fast
X1
Slow
X2
Fwd
X3
Rev
X4

Figure 7.7

Name ______________________________ Date ______________

Lab 

Write Two-Way Light

Write a program to control the brightness of a LED. Using a 12V and a 6V power supply, connect a LED in series with a 1kohm to the PLC. Have X0 turn on the light for high intensity and X1 low intensity. Have the circuit latch and let X2 reset the latch. Refer to Figure 7.7 for help.

Equipment Required

1. PLC
2. 12V and 6V power supply
3. Push button switches
4. 1 kohm resistor
5. LED

Name ______________________________ Date ______________

Lab

Automatic Toilet Seat

A toilet seat is designed to open when sensor X0 is activated. The seat motor turns on through output Y2. When the seat is all the way up, sensor X3 is activated and the motor stops. When the seat is up and X0 is activated, the seat motor turns on but this time the motor is told to go in reverse through output Y3. Y3 controls a relay which changes the polarity on the motor in order to reverse it. When the seat is all the way down, sensor X2 is activated telling the motor to stop. Refer to Figure 7.8.

To recap:

X0—Activate the seat motor to turn on

X3—Seat up sensor

X2—Seat down sensor

Y2—Turn on motor

Y3—Reverse motor

Equipment Required

1. PLC
2. 12V power supply
3. 12V motor
4. Push button switches
5. DPDT 12V relay

To demonstrate the lab, starting with the seat down, momentarily apply 12VDC to X0, and observe the rotation of the motor. To simulate the seat reaching the up position, apply 12VDC to X3 and keep it there. Next, momentarily apply 12V to X0 while observing the rotation of the motor. Take the 12V off of X3 and momentarily apply 12V to X2, the down position, to stop the motor.

Questions

1. If X0 is activated and the seat is down, what does the seat do?
2. If the seat is down and not moving, are X2 and X3 activated or not?
3. If the seat is moving up, are X2 and X3 activated or not?

4. What turns off the motor when the seat is moving up?
5. What turns off the motor when the seat is moving down?
6. If the motor does not turn off when the seat is moving up, what is the problem? Assume PLC OK and outputs are working.
7. If the motor does not turn off when the seat is moving down, what is the problem? Assume PLC OK and outputs are working.
8. If the seat never goes up, what could be the problem? Assume PLC OK and outputs are working.
9. If the seat never goes down, what could be the problem? Assume PLC OK and outputs are working.

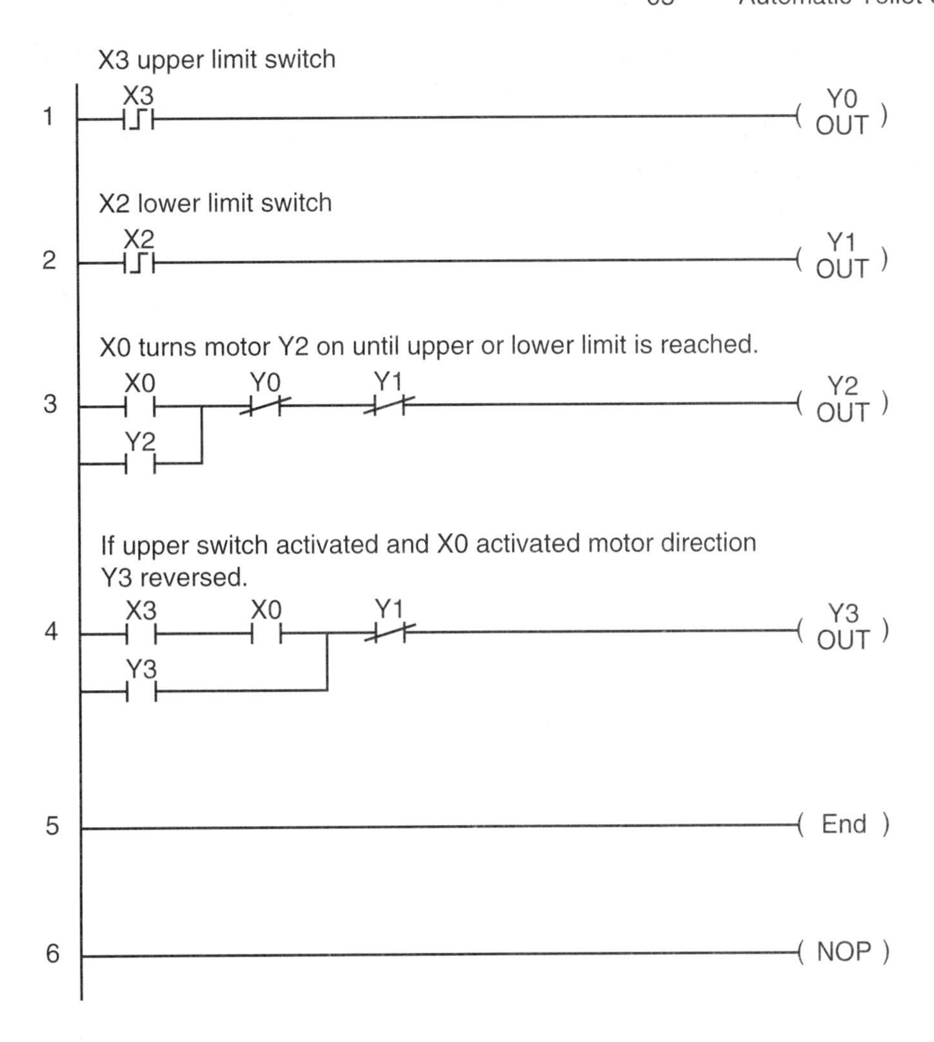
05 Automatic Toilet Seat
X3 upper limit switch
1
X3
Y0
OUT
X2 lower limit switch
2
X2
Y1
OUT
X0 turns motor Y2 on until upper or lower limit is reached.
3
X0
Y0
Y1
Y2
OUT
Y2
If upper switch activated and X0 activated motor direction
Y3 reversed.
4
X3
X0
Y1
Y3
OUT
Y3
5
End
6
NOP

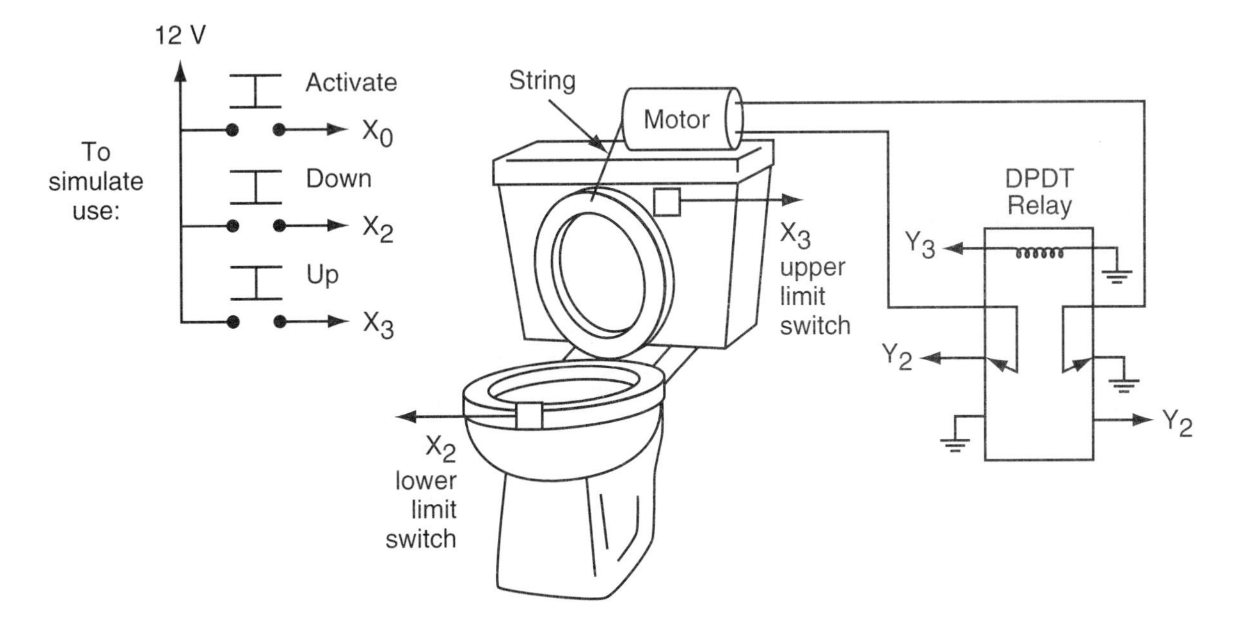
12 V
Activate
X0
To
simulate
use:
Down
X2
Up
X3
String
Motor
X3
upper
limit
switch
DPDT
Relay
Y3
Y2
Y2
X2
lower
limit
switch

Figure 7.8

Name ______________________________ Date ______________

Lab 

Shade-O-Matic

A shade is designed to move up and down with light level: up when the sensor is lit and down when the sensor is covered. A sensor X0 activates the shade. The shade motor turns on through output Y2. When the shade is all the way up, sensor X3 is activated and the motor stops. When the shade is up and X0 is activated, the shade motor turns on but this time the motor is told to go in reverse through output Y3. Y3 controls a relay which changes the polarity on the motor in order to reverse it. The output of the photo resistor voltage divider circuit outputs a low voltage when the sensor is dark and a higher voltage when it is lit. Refer to Figure 7.9.

To recap:

X0—Activate the shade motor to turn on

X3—Shade up sensor

X2—Shade down sensor

Y2—Turn on motor

Y3—Reverse motor

Equipment Required

1. PLC
2. 12V power supply
3. 12V motor
4. Photo resistor
5. 10kohm pot
6. DPDT 12V relay
7. 1kohm resistor

To demonstrate the lab, start by adjusting the 10kohm pot so that X0 is activated with a covering or uncovering of the photo resistor. Starting with the shade down, momentarily apply light to the sensor. Observe the rotation of the motor. To simulate the shade reaching the up position, momentarily apply 12V to X3. The motor should stop. Next, cover the sensor so that a high to a low transition occurs on X0, while observing the rotation of the motor. Simulate the shade reaching the down position by momentarily applying 12V to X2. The motor should stop.

Questions

1. If X0 is activated and the shade is down, what does the shade do?
2. If the shade is down, are X2 and X3 activated or not?
3. If the shade is up, are X2 and X3 activated or not?
4. What turns off the motor when the shade is moving up?
5. What turns off the motor when the shade is moving down?
6. If the motor does not turn off when the shade is moving up, what is the problem?
7. If the motor does not turn off when the shade is moving down, what is the problem?
8. If the shade never goes up, what could be the problem?
9. If the shade never goes down, what could be the problem?

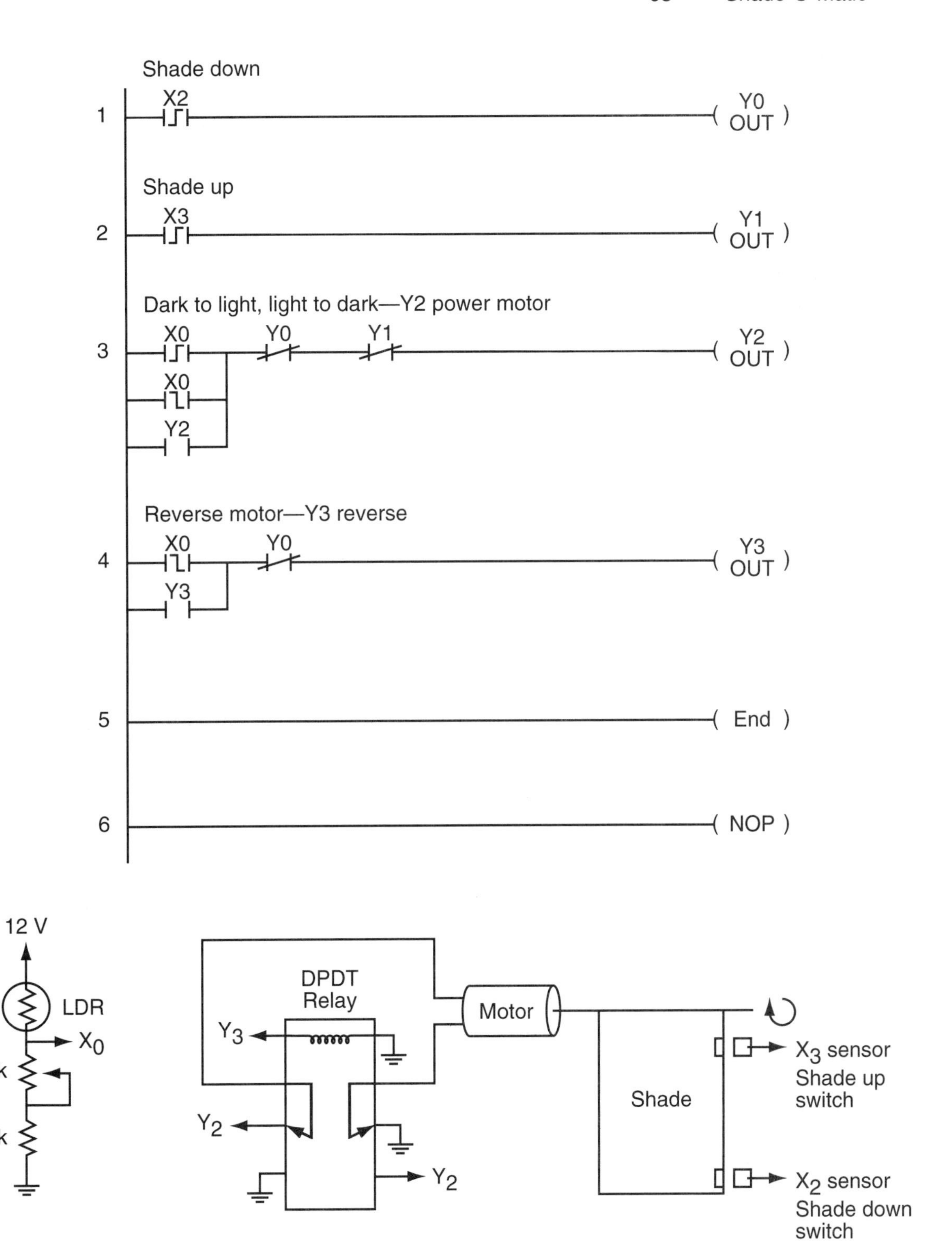
05 Shade-O-Matic
Shade down
X2
Y0 OUT
1
Shade up
X3
Y1 OUT
2
Dark to light, light to dark—Y2 power motor
X0
Y0
Y1
X0
Y2
Y2 OUT
3
Reverse motor—Y3 reverse
X0
Y0
Y3
Y3 OUT
4
5
End
6
NOP
12 V
LDR
X0
10k
1k
DPDT
Relay
Y3
Y2
Y2
Motor
Shade
X3 sensor
Shade up switch
X2 sensor
Shade down switch

Figure 7.9

Name ______________________________ Date ______________

Lab 46

Draw Bridge

A draw bridge is designed to open when a boat passes sensor X0. The bridge motor turns on through output Y2. When the bridge is all the way up, sensor X3 is activated and the motor stops. The boat passes under the bridge and activates sensor X1, which then turns on the bridge motor again but this time the motor is told to go in reverse through output Y3. Y3 controls a relay which changes the polarity on the motor in order to reverse it. When the bridge is all the way down, sensor X2 is activated and stops the motor. Refer to Figure 7.10.

To recap:

X0—Activate the bridge motor to move up

X1—Activate the bridge motor to move down

X3—Bridge up sensor

X2—Bridge down sensor

Y2—Turn on motor

Y3—Reverse motor

Equipment Required

1. PLC
2. 12V power supply
3. 12V motor
4. DPDT 12V relay
5. Momentary switches

To demonstrate the lab, momentarily apply 12V to X0. Observe the rotation of the motor. To simulate the bridge reaching the up position, momentarily apply 12V to X3. The motor should stop. Next, momentarily apply 12V to X1 and observe the rotation of the motor. Simulate the bridge reaching the down position by momentarily applying 12V to X2. The motor should stop.

Questions

1. If X0 is activated, what does the bridge do?
2. If the bridge is down, are X2 and X3 activated or not?

3. If the bridge is up, are X2 and X3 activated or not?
4. What turns off the motor when the bridge is moving up?
5. What turns off the motor when the bridge is moving down?
6. If the motor does not turn off when the bridge is moving up, what is the problem?
7. If the motor does not turn off when the bridge is moving down, what is the problem?
8. If the bridge never goes up, what could be the problem?
9. If the bridge never goes down, what could be the problem?

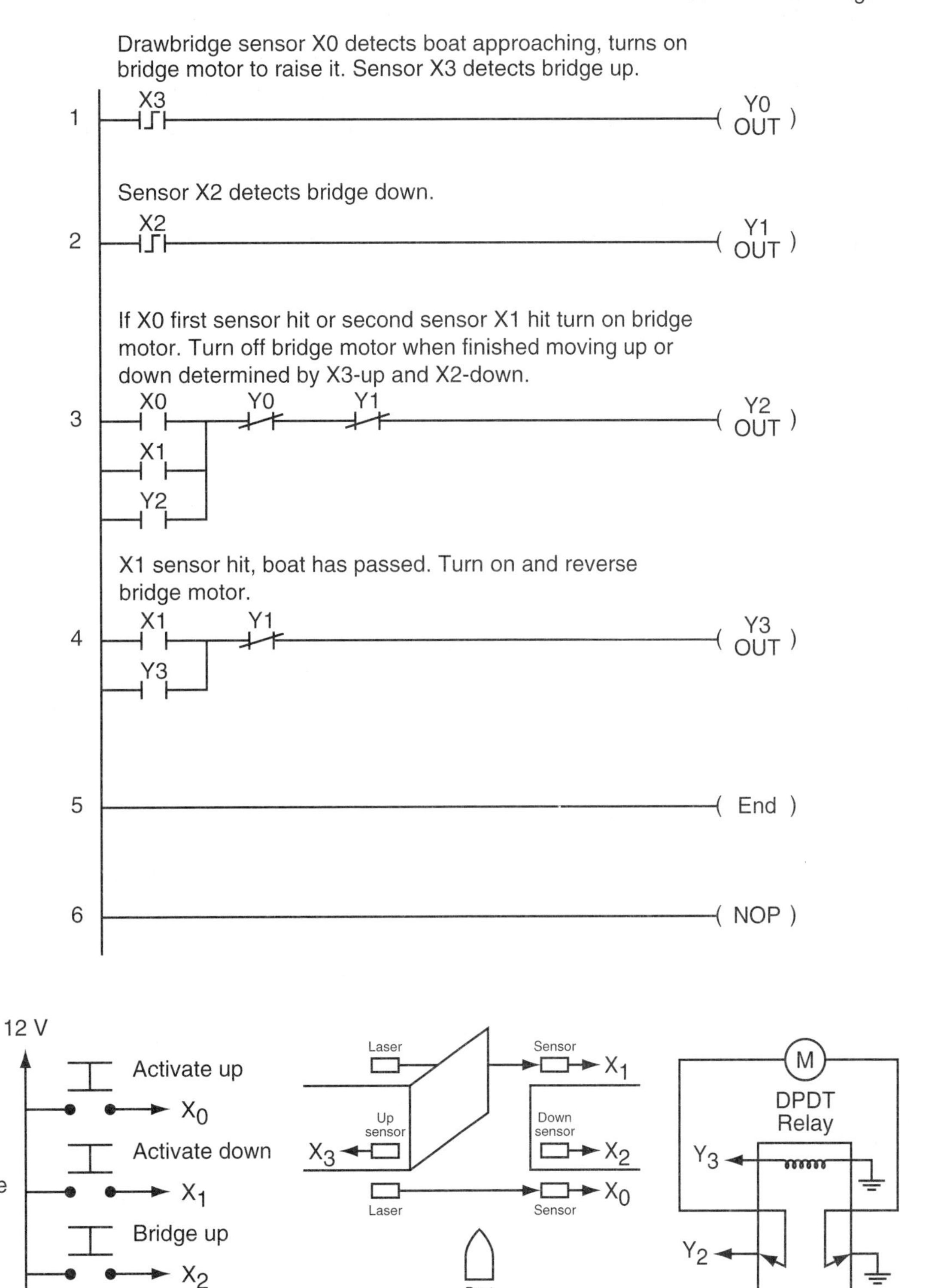
05 Drawbridge
Drawbridge sensor X0 detects boat approaching, turns on bridge motor to raise it. Sensor X3 detects bridge up.
1
X3
Y0 OUT
Sensor X2 detects bridge down.
2
X2
Y1 OUT
If X0 first sensor hit or second sensor X1 hit turn on bridge motor. Turn off bridge motor when finished moving up or down determined by X3-up and X2-down.
3
X0
Y0
Y1
X1
Y2
Y2 OUT
X1 sensor hit, boat has passed. Turn on and reverse bridge motor.
4
X1
Y1
Y3
Y3 OUT
5
End
6
NOP
12 V
To simulate use:
Activate up
X0
Activate down
X1
Bridge up
X2
Bridge down
X3
Laser
Sensor
X1
Up sensor
X3
Down sensor
X2
Laser
Sensor
X0
Boat
M
DPDT Relay
Y3
Y2
Y2

Figure 7.10

Name ______________________________ Date ______________

Lab

People Counter

The people counter is designed to prevent a room from ever exceeding occupancy capacity by raising and lowering a gate permitting or barring entrance. Two sensors are placed close together at the entrance determining whether someone is entering or leaving, incrementing or decrementing an up/down counter. Starting with the gate up when a count of 5 is reached, the gate will move down by turning on output Y2 (power) and output Y3 (direction).Y3 controls a relay which changes the polarity on the motor in order to reverse it. When the gate is all the way down, X2 tells the motor to stop. When the gate is all the way up, sensor X3 tells the motor to stop. The gate is designed so that it can be manually opened from the inside to permit exit. If the count drops below 5 because of people leaving, the gate will move up and stop when the up sensor, X3, is activated. Refer to Figure 7.11.

To recap:

X0, X1–Activate the gate motor to turn on

X3—Gate up sensor

X2—Gate down sensor

Y2—Turn on motor

Y3—Reverse motor

Equipment Required

1. PLC
2. 12V power supply
3. 12V motor
4. Push button switches
5. DPDT 12V relay

To demonstrate the lab, start with a count of zero by applying 12V to X4 to reset the counter. Apply 12V and hold it on X0, then apply 12V to X1 momentarily. This simulates someone passing through the beams entering the room. The beams are placed close together so that both will be broken at same time during passage. Entering is counted differently than exiting: when entering, X0 will be activated first then X1; when someone exits, X1 will first be activated then X0. Simulate 5 people entering and observe the output. The gate will close when a count of 5 is reached and stop when sensor X2 is activated. Simulate someone leaving by first breaking X1's beam then X0's by applying 12V to X1, holding it active, then activating X0 momentarily. The count will decrement and the gate will open and stop when the up sensor, X3, is activated.

Questions

1. If X0 is activated and then X1 is activated and the gate is down, what does the gate do?
2. If the gate is down, are X2 and X3 activated or not?
3. If the gate is up, are X2 and X3, activated or not?
4. What turns off the motor when the gate is moving up?
5. What turns off the motor when the gate is moving down?
6. If the motor does not turn off when the gate is moving up, what is the problem?
7. If the motor does not turn off when the gate is moving down, what is the problem?
8. If the gate never goes up, what could be the problem?
9. If the gate never goes down, what could be the problem?

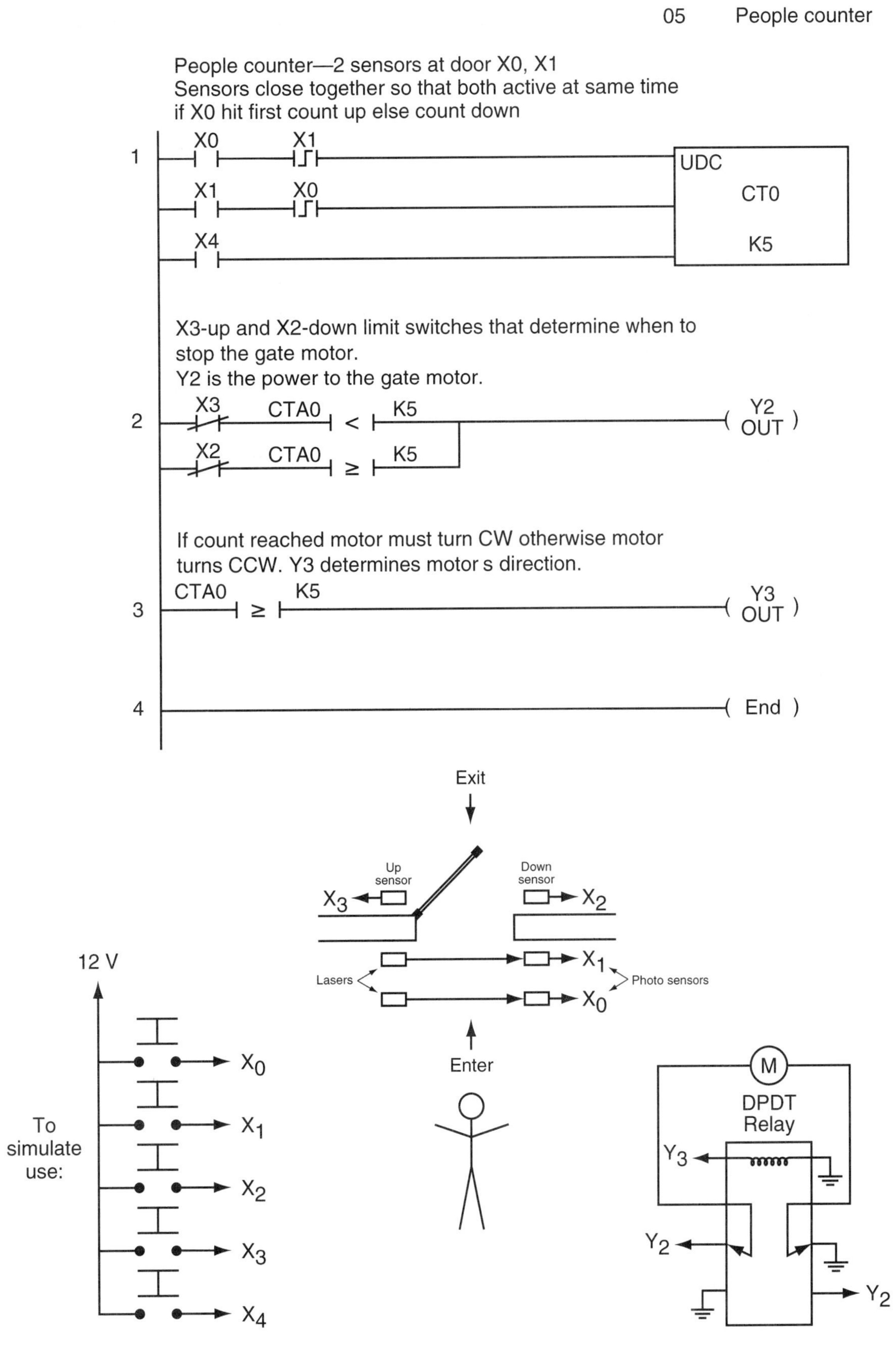

05 People counter

People counter—2 sensors at door X0, X1
Sensors close together so that both active at same time
if X0 hit first count up else count down

1 X0 X1 — UDC CT0 K5
X1 X0
X4

X3-up and X2-down limit switches that determine when to stop the gate motor.
Y2 is the power to the gate motor.

2 X3 CTA0 < K5 — Y2 OUT
X2 CTA0 ≥ K5

If count reached motor must turn CW otherwise motor turns CCW. Y3 determines motor s direction.

3 CTA0 ≥ K5 — Y3 OUT

4 — End

Exit
Up sensor
Down sensor
X_3
X_2
X_1
X_0
Lasers
Photo sensors
Enter

12 V
To simulate use:
X_0
X_1
X_2
X_3
X_4

M
DPDT Relay
Y_3
Y_2
Y_2

Figure 7.11

Name ______________________ Date ____________

Encoder FWD REV

This program is designed to control the direction and angular position of a motor's shaft. The shaft angle is determined by an incremental encoder that puts out one pulse for every degree it turns. A 741 op-amp is used as a buffer between the encoder and the PLC to prevent loading of the encoder by the PLC. The program is designed to rotate the motor's shaft 360 deg and stop. X0 starts the motor, X1 stops the motor, X2 selects reverse, and X3 selects forward. The output from the encoder is connected to the 741 buffer, then to X4. X5 resets the counter. Y0 controls the motor's power and Y1 controls the motor's direction. Refer to Figure 7.12.

To recap:

X0—Start
X1—Stop
X2—Reverse
X3—Forward
X4—Encoder
X5—Reset counter
Y0—Motor power
Y1—Motor's direction

Equipment Required

1. PLC
2. 12V power supply
3. 12V motor
4. Push button switches
5. 10kohm resistor
6. DPDT 12V relay
7. 741 op-amp
8. Incremental encoder

To demonstrate the lab, momentarily apply 12V to X5 to reset the counter. Momentarily apply 12V to X0 and observe the rotation of the motor. Observe the count on the counter.

When the count on the counter reaches 360, the motor should stop. If the motor comes to a stop for a count a little greater than 360, it is because there is some momentum in the motor, and it spins a little after the power has been turned off. Reset the counter again, then momentarily apply 12V to X2 for reverse, then 12V to X0. Observe the motor run in reverse and stop at a count of 360.

Questions

1. What happens if X0 is selected?
2. What happens if X1 is selected?
3. What happens if X2 is selected?
4. What happens if the motor is running in the forward direction and the reverse button is hit?
5. What parameter needs to be changed to make the shaft rotate 2 revolutions. What is it changed to?
6. What could be the problem if the motor will not go into reverse?

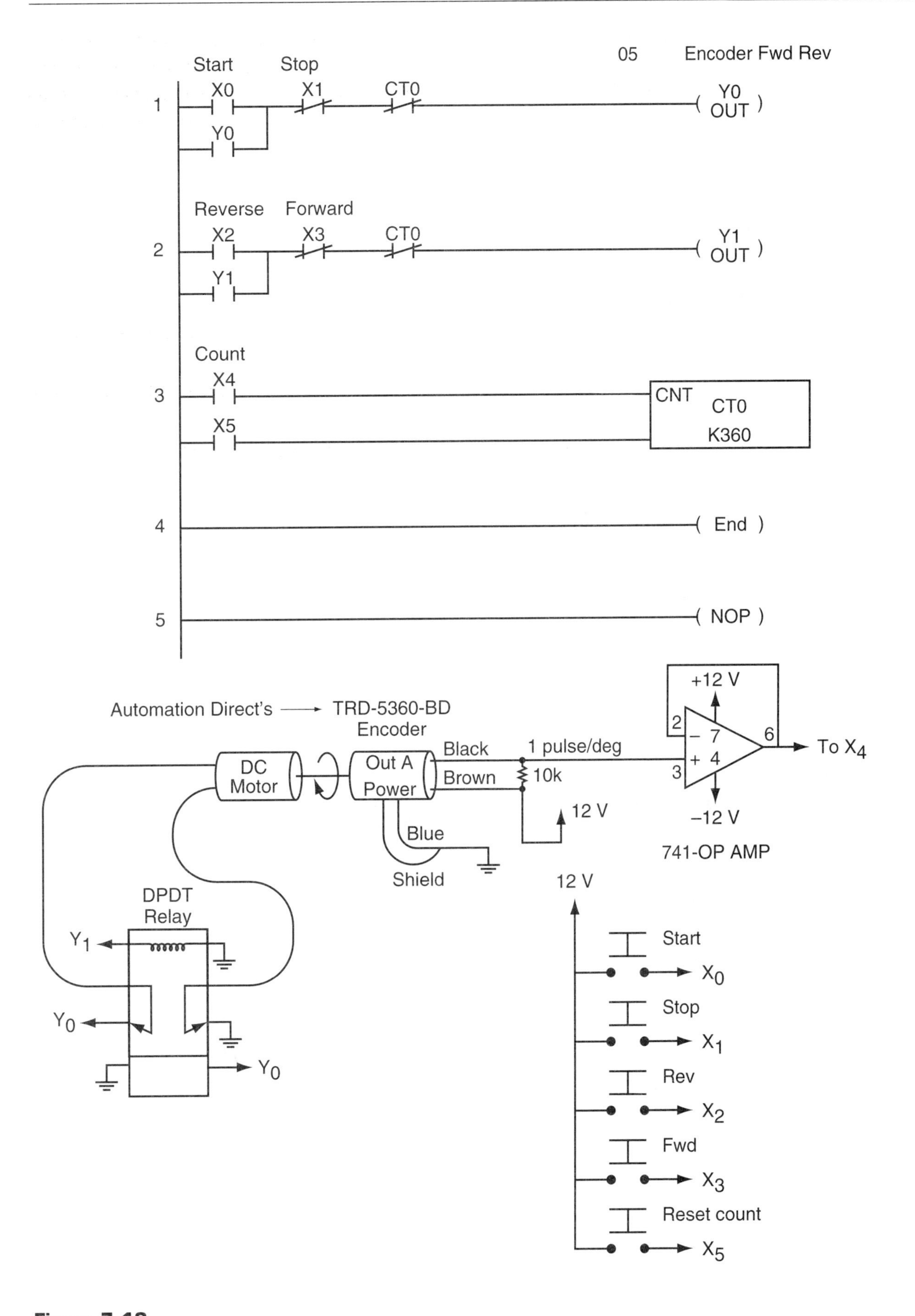

Start
Stop
05
Encoder Fwd Rev
X0
X1
CT0
Y0 OUT
Y0
1
Reverse
Forward
X2
X3
CT0
Y1 OUT
Y1
2
Count
X4
3
CNT
CT0
K360
X5
4
End
5
NOP
Automation Direct's
TRD-5360-BD Encoder
DC Motor
Out A
Power
Black
Brown
Blue
Shield
1 pulse/deg
10k
12 V
+12 V
−12 V
741-OP AMP
To X4
DPDT Relay
Y1
Y0
12 V
Start
X0
Stop
X1
Rev
X2
Fwd
X3
Reset count
X5

Figure 7.12

Name ______________________ Date ______________

Lab

Elevator

A 3-floor elevator is controlled by selecting one of 3 floor switches. It moves to the 3rd floor when X3 is selected, to the 2nd floor if X2 is selected, and the 1st floor if X1 is selected. The elevator stops at each of the selected floors when a limit switch for the selected floor is activated. The elevator's motor is controlled by Y2, and is reversed (through Y3) if the elevator is directed to move to a lower floor than it is currently on. Refer to Figure 7.13.

To recap:

X1—1st floor
X2—2nd floor
X3—3rd floor
X4—1st floor limit switch
X5—2nd floor limit switch
X6—3rd floor limit switch
Y2—Motor power
Y3—Reverse motor

Equipment Required

1. PLC
2. 12V power supply
3. 12V motor
4. Push button switches
5. DPDT 12V relay

To demonstrate the lab, start with the elevator on the first floor by activating X4, the first floor switch. With X4 activated, select the second floor by momentarily activating X2, and observe the rotation of the motor. When the motor starts, deactivate X4 and simulate the elevator reaching the second floor by activating X5. The motor should stop. Next, with X5 activated, select the third floor by momentarily applying 12V to X3, while observing the rotation of the motor. Deactivate X5, and simulate the elevator reaching the third floor by activating switch X6. With X6 activated, have the elevator go back down to the second floor by activating X2. The motor should be moving in reverse now. Deactivate X6 and simulate the elevator reaching the second floor with switch X5.

Questions

1. What happens if the elevator is on the 2nd floor and X3 is pressed?
2. What shuts off the motor when the 3rd floor is reached?
3. What shuts off the motor when the selected floor is reached?
4. If the motor never shuts off when the selected floor is reached, what could be the problem?
5. If the elevator only moves upward and never downward, what could be the problem?
6. If the elevator will not move to the second floor, what could be the problem?
7. What is the purpose of the relay in the circuit diagram?

05 Elevator

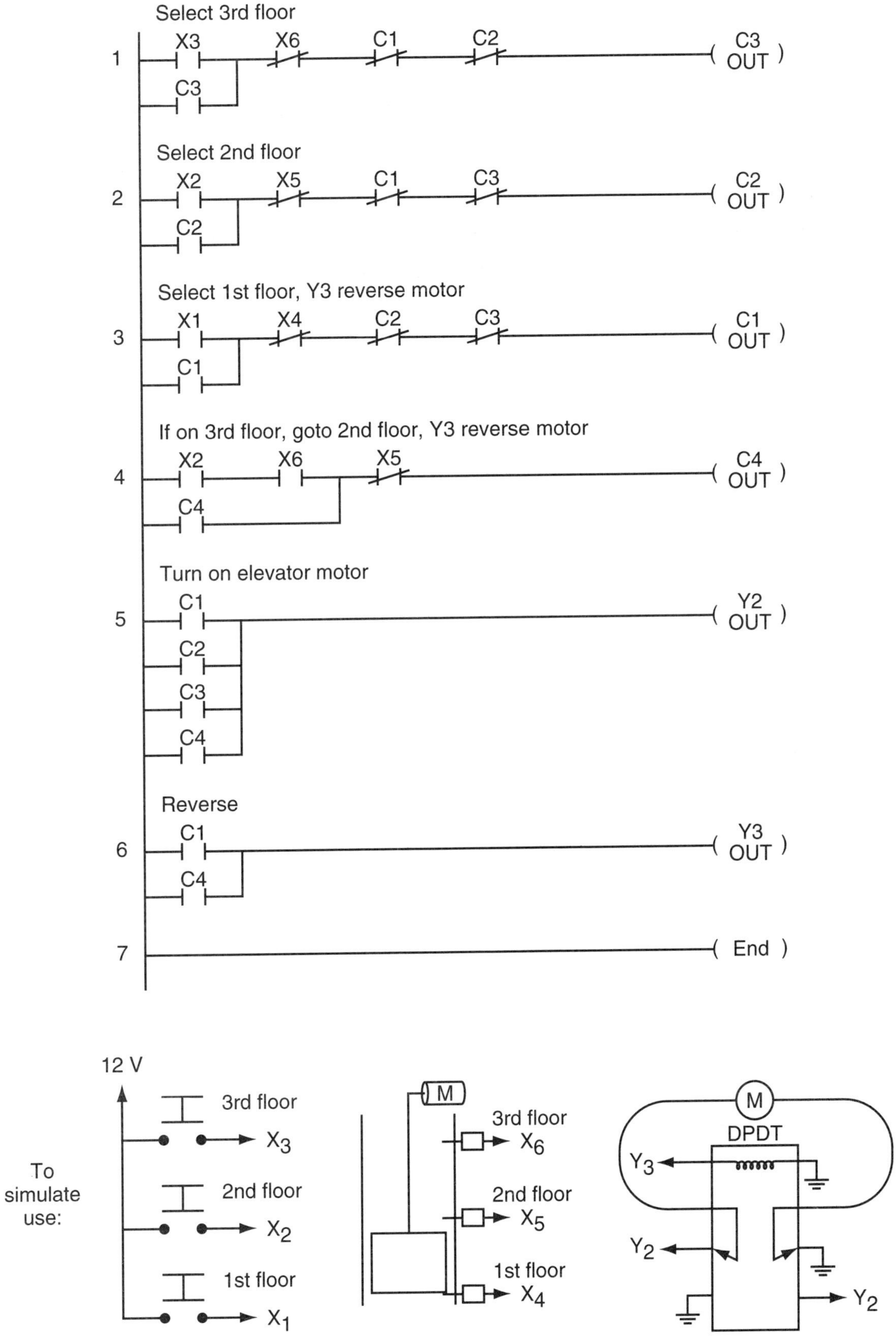

Select 3rd floor
1 X3 X6 C1 C2 (C3 OUT)
C3

Select 2nd floor
2 X2 X5 C1 C3 (C2 OUT)
C2

Select 1st floor, Y3 reverse motor
3 X1 X4 C2 C3 (C1 OUT)
C1

If on 3rd floor, goto 2nd floor, Y3 reverse motor
4 X2 X6 X5 (C4 OUT)
C4

Turn on elevator motor
5 C1 (Y2 OUT)
C2
C3
C4

Reverse
6 C1 (Y3 OUT)
C4

7 (End)

12 V
3rd floor X_3
2nd floor X_2
1st floor X_1
To simulate use:

M
3rd floor X_6
2nd floor X_5
1st floor X_4

M
DPDT
Y_3
Y_2
Y_2

Figure 7.13

SECTION 8

PLCs with Analog to Digital and Digital to Analog Converters

Objectives

Upon completion of these labs, you should be able to:

- Understand how Analog to Digital (ADC) and Digital to Analog Converters (DAC) are used with PLCs to acquire data, process it, and output it.
- Understand how ON/OFF control works.
- Understand how proportional control works.
- Understand how time proportioning control works.

Introduction

The following labs illustrate the use of PLCs with Digital to Analog Converters, DACs, and Analog to Digital Converters, ADCs. Connections with analog sensors are now possible along with on/off and proportional control. Because of the wide variety of PLCs, ADCs, and DACs, specific wiring and labels (inputs, outputs, and commons) vary among them. Be sure to consult your spec sheet for your particular PLC, ADC, and DAC. These labs are specifically related to AutomationDirect's® DL05 PLC with a F0-2AD2DA-2 analog input/output module.

ADC and DAC Review

For a 12 bit ADC/DAC, which we will be using in these labs, there are 0–4095 different states the input or output voltage is divided into. The formulas that describe these are:

$$ADCoutput = \frac{Vin}{Vref} \times 4095$$

$$DACoutput = \frac{DataOut}{4095} \times Vref$$

The reference voltage used in the following labs is 10V.

In the following programs, the PLC needs to be told where incoming data from the ADC is stored and where to send the outgoing data to the DAC. The PLC needs only to be told once, so a special relay contact SP0 is used which closes only once on the first scan on the program. See Figure 8.1.

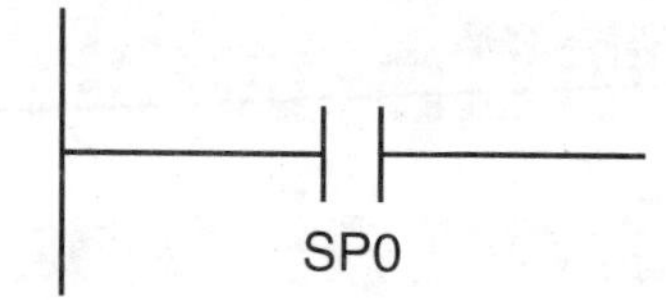

Figure 8.1 Special relay SPO closes only once on the first scan of the program.

Addresses and data first have to be written to the accumulator in the PLC before they can be sent to the appropriate location. The following commands shown in Figures 8.2–8.8 will be used in setting up the PLC to use an ADC/DAC.

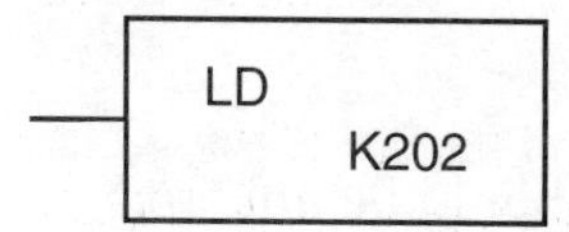

Figure 8.2 LD: The LD instruction loads the accumulator with a constant number or a memory location.

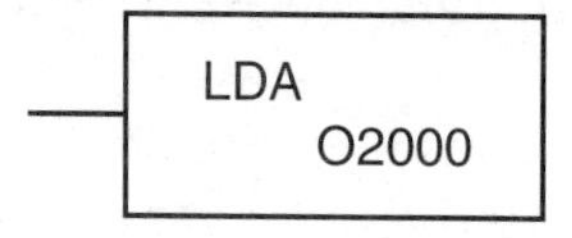

Figure 8.3 LDA: The LDA instruction converts an octal number into hex and loads it into the accumulator.

Figure 8.4 OUT: The OUT instruction copies the value in the accumulator to a specified memory location.

Addition, subtraction, multiplication, and division can be performed by the PLC. See Figures 8.5–8.8.

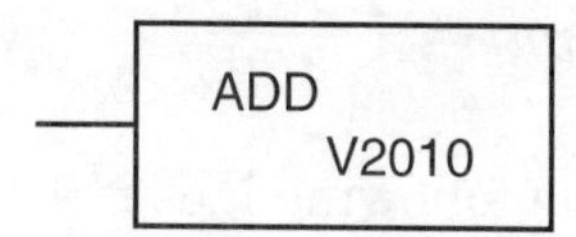

Figure 8.5 ADD: The ADD instruction adds the number in a specified memory location with the value in the accumulator and stores the result in the accumulator.

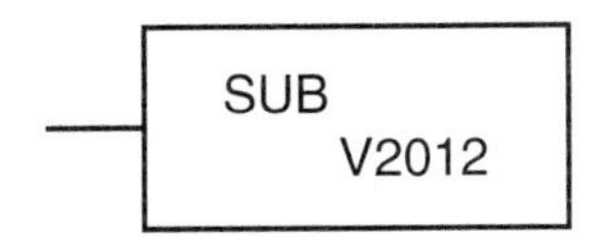

Figure 8.6 SUB: The SUB instruction subtracts the number in a specified memory location from the value in the accumulator and stores the result in the accumulator.

Figure 8.7 MUL: The MUL instruction multiplies the number in a specified memory location from the value in the accumulator and stores the result in the accumulator.

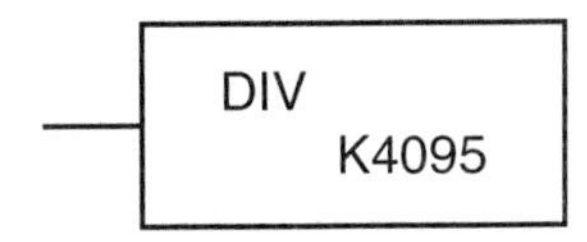

Figure 8.8 DIV: The DIV instruction divides the value in the accumulator by the number in a specified memory location and stores the result in the accumulator.

Name ______________________________ Date ______________

On-Off Temperature Control Using a Fan and an ADC

This lab demonstrates on/off temperature control. The temperature of a room is controlled with a fan. The temperature is measured with an LM34 temperature sensor, which puts out 10mV/°F. Consult the spec sheet in Appendix B, Figure B.4, for more details on the LM34. The LM34 is connected to channel 2 of the ADC and a 10K pot to channel 1 of the ADC, which acts as the set point temperature. When the temperature rises above the set point the fan, which is connected to Y1, turns on. A flow chart diagram illustrating the PLC program showing the function of each rung is shown in Figure 8.9. The wiring diagram and the program are illustrated in Figures 8.10 and 8.11.

To recap:

Input Channel 1—Input set point

Input Channel 2—Input temperature sensor

Y1—Output to fan

Equipment Required

1. PLC with an ADC
2. 12V, 5V, and 24V power supply
3. LM34 temperature sensor
4. 12V motor

To demonstrate the lab, wire the circuit in Figure 8.10, and type in the program in Figure 8.11.

View the set point from channel 1 stored at V2010, and the temperature from channel 2 stored at V2012 by following these steps:

1. Click on **Debug** on the tool bar.
2. Click on **Data View** and type in the names of the variables you would like to see, that is, V2010 and V2012.
3. Click on **Status** on the tool bar to see the current value of the variables listed.

Adjust the pot until the set point temperature is a little above the temperature the LM34 is reading. Place your fingers on the LM34, raising its temperature above the set point. At this point the fan should turn on.

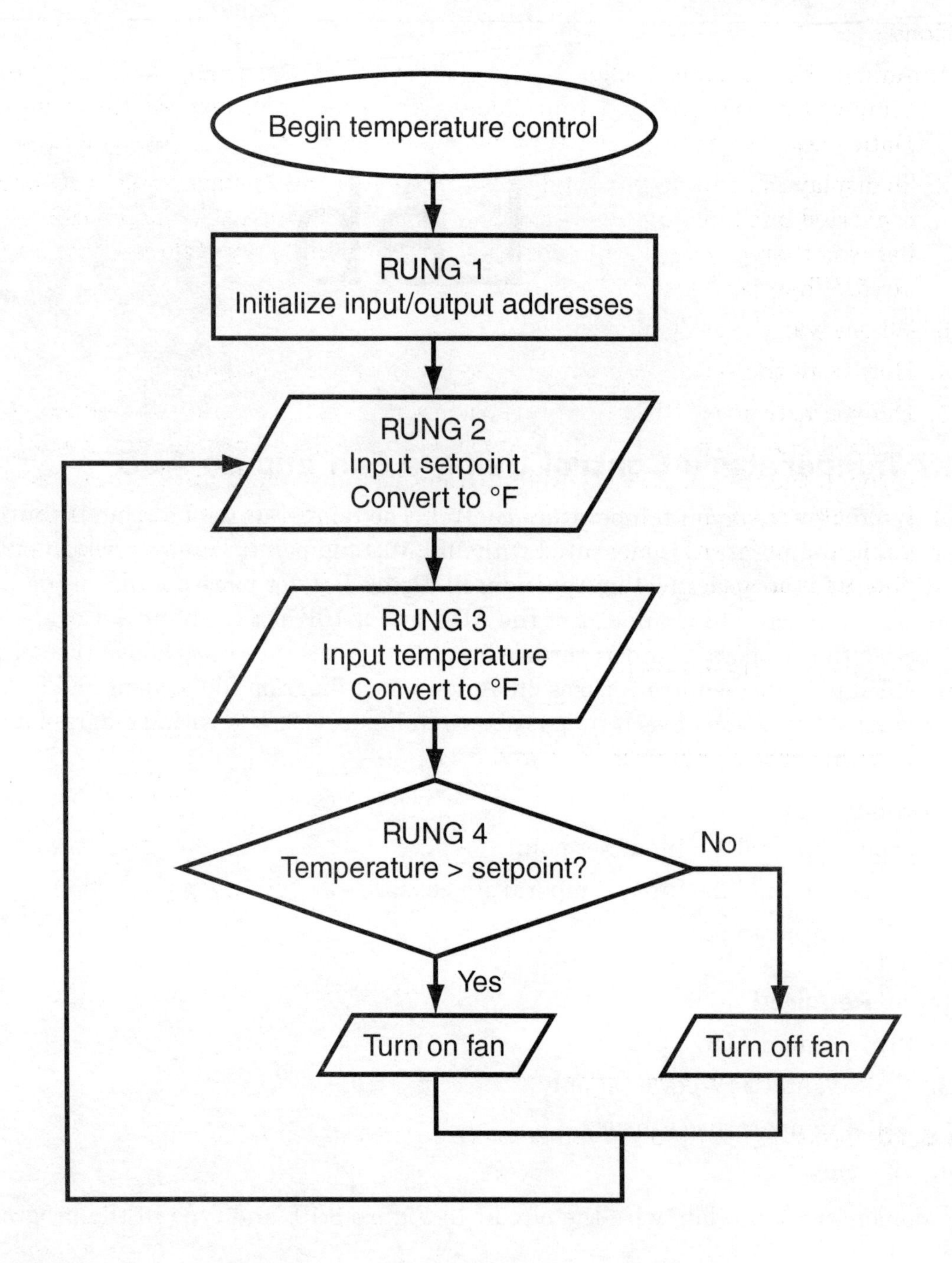

Figure 8.9 Flow chart diagram of on-off temperature control program.

Questions

1. Adjust the set point to slightly above room temperature, say 75°F. Now heat the temperature sensor with your fingers and watch the temperature rise in the Data View. At what temperature did the fan come on?
2. To display the data in term of degrees Fahrenheit, the voltage input, 10mV/°F, was converted into a binary number from a 12 bit ADC with a 10V reference. This number was then converted into degrees by multiplying it by 1000 and dividing it by 4095. Where was this temperature stored?
3. Where was the set point stored?
4. How were the set point and the actual temperature compared?
5. Did you notice how the output turned on and off intermittently as the temperature reached the set point? This is because the temperature did not steadily pass through the set point but moved up and down through it until it was consistently greater than the set point. To eliminate this problem we can have the turn on point at a different temperature than the turn off point. This is called hysteresis, and is illustrated in Lab 51.

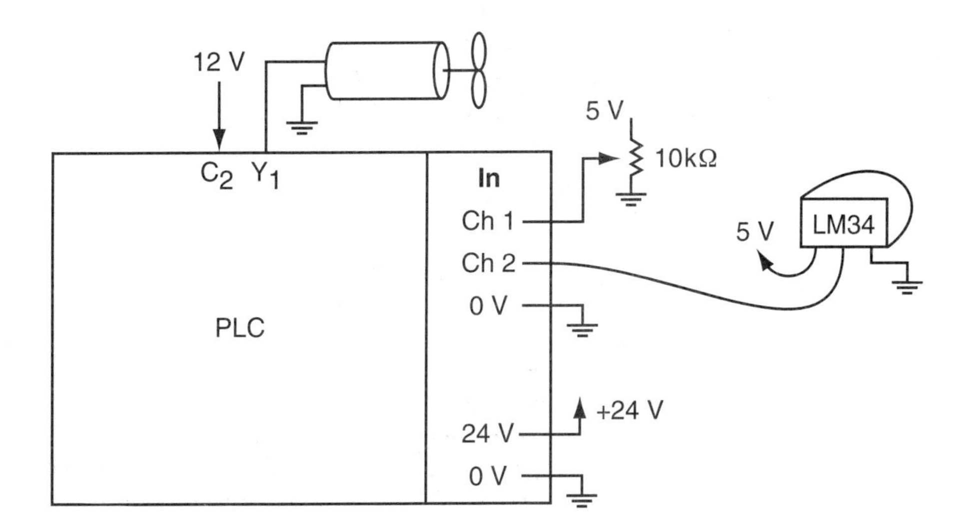

Figure 8.10 Circuit layout for on-off temperature control with an ADC.

05 On/Off Temp Control with ADC

On/off temperature control program using an LM34. Set point channel 1, process variable channel 2

Program turns on a fan at Y1 if the temperature is greater than the set point.

SP0 closed on 1st scan.
LD K202, out V7700—scan 2 channels in and 2 out, BCD format

LDA O2000, out 7701—loads incoming data s memory location
Ch1 in - V2000, Ch2 in - V2001

LDA O2010, out 7702—loads outgoing data s memory location
Ch1 out - V2010, Ch2 out - V2011

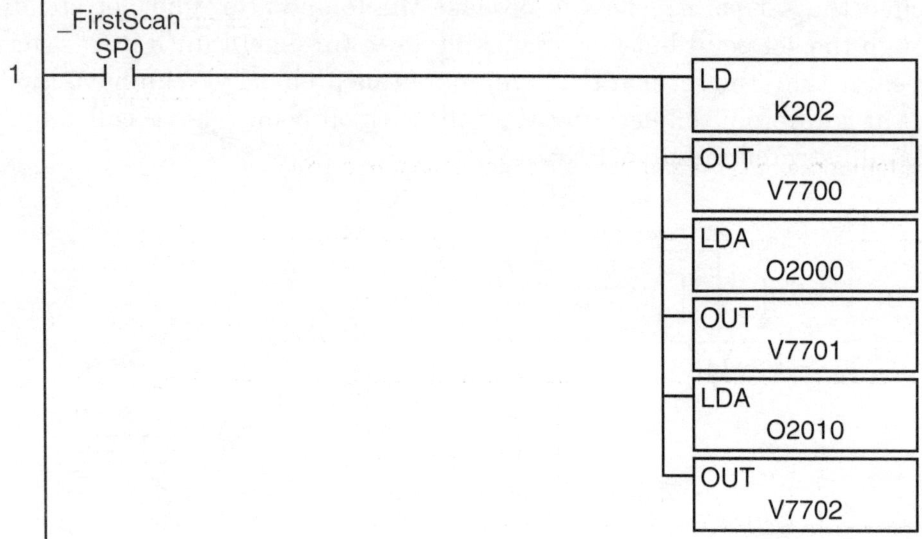

_FirstScan
SP0

1

LD K202
OUT V7700
LDA O2000
OUT V7701
LDA O2010
OUT V7702

Math used to convert input into degrees F.
Load accumulator with channel 1 set point stored at V2000.
Multiply and divide and store at V2010.

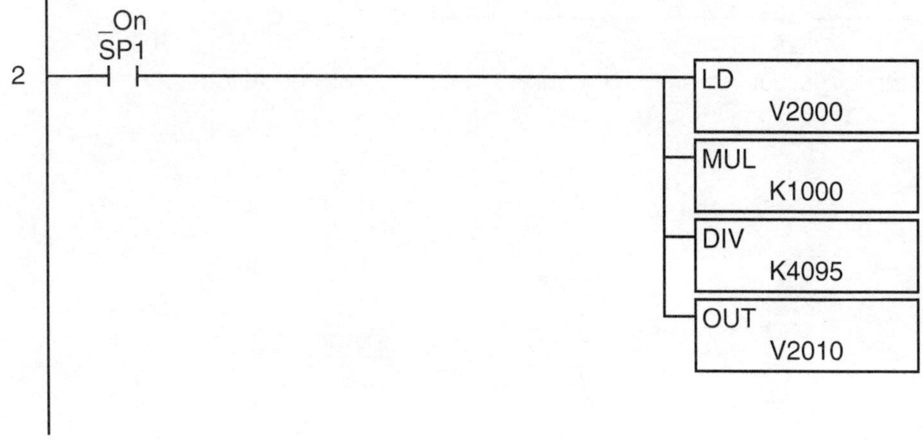

_On
SP1

2

LD V2000
MUL K1000
DIV K4095
OUT V2010

Figure 8.11a

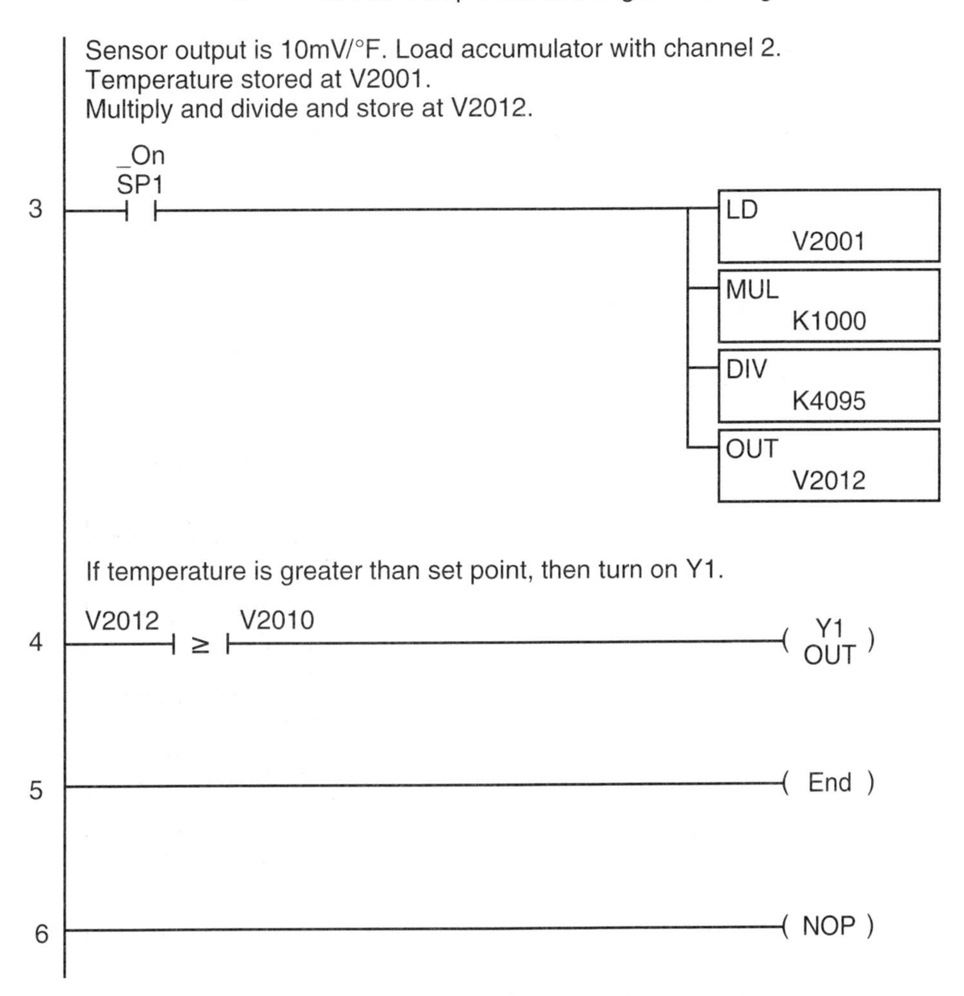
05 On/Off Temp Control Using a Soldering Iron and an ADC
Sensor output is 10mV/°F. Load accumulator with channel 2.
Temperature stored at V2001.
Multiply and divide and store at V2012.
_On
SP1
3
LD
V2001
MUL
K1000
DIV
K4095
OUT
V2012
If temperature is greater than set point, then turn on Y1.
4
V2012
≥
V2010
Y1
OUT
5
End
6
NOP

Figure 8.11b

Name ______________________________ Date ______________

Lab

On-Off Temperature Control Using a Soldering Iron and an ADC

This lab demonstrates temperature control of a soldering iron. The temperature of the iron is controlled by turning the iron on when it is below the set point, and off when it is above the set point. The temperature is measured with an LM34 temperature sensor, which puts out 10mV/°F. It is connected to channel 2 of the ADC and a 10kohm pot to channel 1 of the ADC, which acts as the set point temperature. The iron's tip is placed into a metal block along with the temperature sensor. When the temperature falls below the set point the iron is turned on, which is connected to Y1. To eliminate the output from intermittently turning on and off as the temperature passes through the set point, hysteresis has been added. That is, the output will turn on at the set point temperature, and will turn off at 2 degrees above the set point. A flow chart diagram illustrating the PLC program showing the function of each rung is shown in Figure 8.12. The wiring diagram and the program are illustrated in Figures 8.13 and 8.14.

To recap:

Input Channel 1—Input set point

Input Channel 2—Input temperature sensor

Y1—Output to soldering iron

Equipment Required

1. PLC with an ADC
2. 12V, 5V, and 24V power supply
3. LM34 temperature sensor
4. Soldering iron
5. Modified extension cord
6. 10K pot
7. A piece of metal to act as a thermal load

To demonstrate the lab, connect up the circuit in Figure 8.13 and type in the program in Figure 8.14.

View the set point from channel 1 stored at V2010, and the temperature from channel 2 stored at V2012 by following these steps:

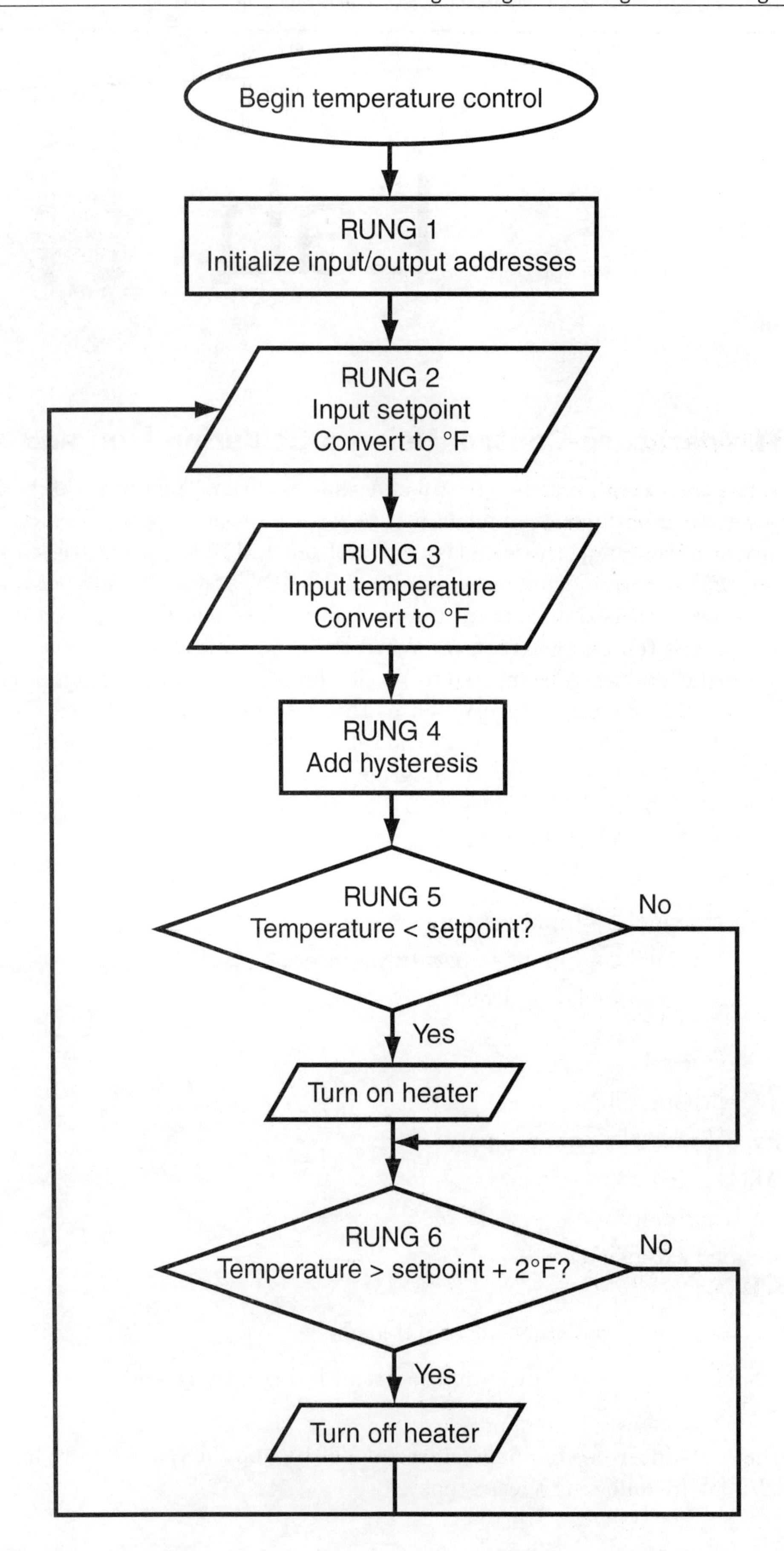

Figure 8.12 Flow chart diagram of on off temperature control using a soldering iron and an ADC.

1. Click on **Debug** on the tool bar.
2. Click on **Data View** and type in the names of the variables you would like to see, that is V2010 and V2012.
3. Click on **Status** on the tool bar to see the current value of the variables listed.

Adjust the pot until the set point temperature is a little below the temperature the LM34 is reading. At this point the soldering iron should turn on, raising the temperature. At a temperature above the set point the iron will turn off.

Questions

1. Adjust the set point to slightly above room temperature, say 100°F. At what temperature did the soldering iron come on?
2. To display the data in terms of degrees Fahrenheit, the voltage input, 10mV/°F, was converted into a binary number from a 12 bit ADC with a 10V reference. This number was then converted into degrees by multiplying it by 1000 and dividing it by 4095. Where was this temperature stored?
3. Where was the set point stored?
4. How were the set point and the actual temperature compared?
5. What is the maximum power soldering iron the PLC can handle? Look at the specs on the PLC.

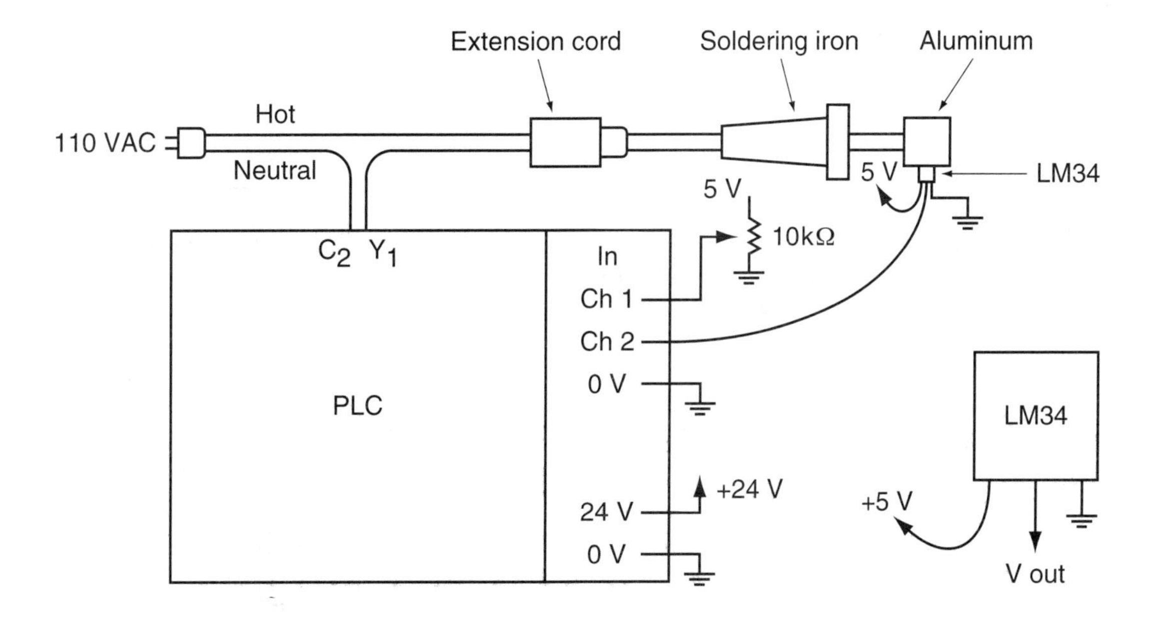

Figure. 8.13 Circuit layout for on-off temperature control using a soldering iron and an ADC.

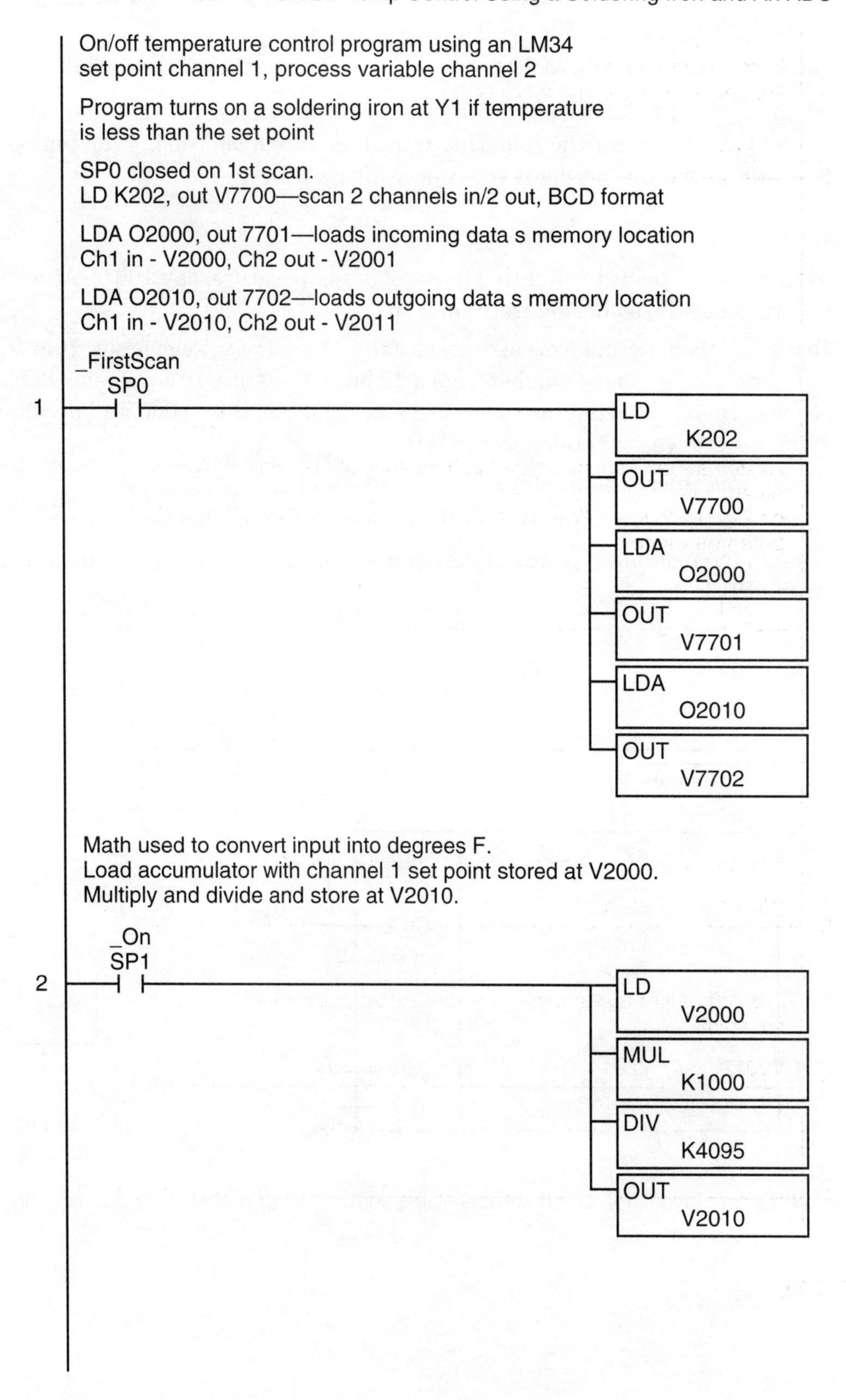
05 On/Off Temp Control Using a Soldering Iron and An ADC
On/off temperature control program using an LM34
set point channel 1, process variable channel 2
Program turns on a soldering iron at Y1 if temperature
is less than the set point
SP0 closed on 1st scan.
LD K202, out V7700—scan 2 channels in/2 out, BCD format
LDA O2000, out 7701—loads incoming data s memory location
Ch1 in - V2000, Ch2 out - V2001
LDA O2010, out 7702—loads outgoing data s memory location
Ch1 in - V2010, Ch2 out - V2011
_FirstScan
SP0
1
LD
K202
OUT
V7700
LDA
O2000
OUT
V7701
LDA
O2010
OUT
V7702
Math used to convert input into degrees F.
Load accumulator with channel 1 set point stored at V2000.
Multiply and divide and store at V2010.
_On
SP1
2
LD
V2000
MUL
K1000
DIV
K4095
OUT
V2010

Figure 8.14a

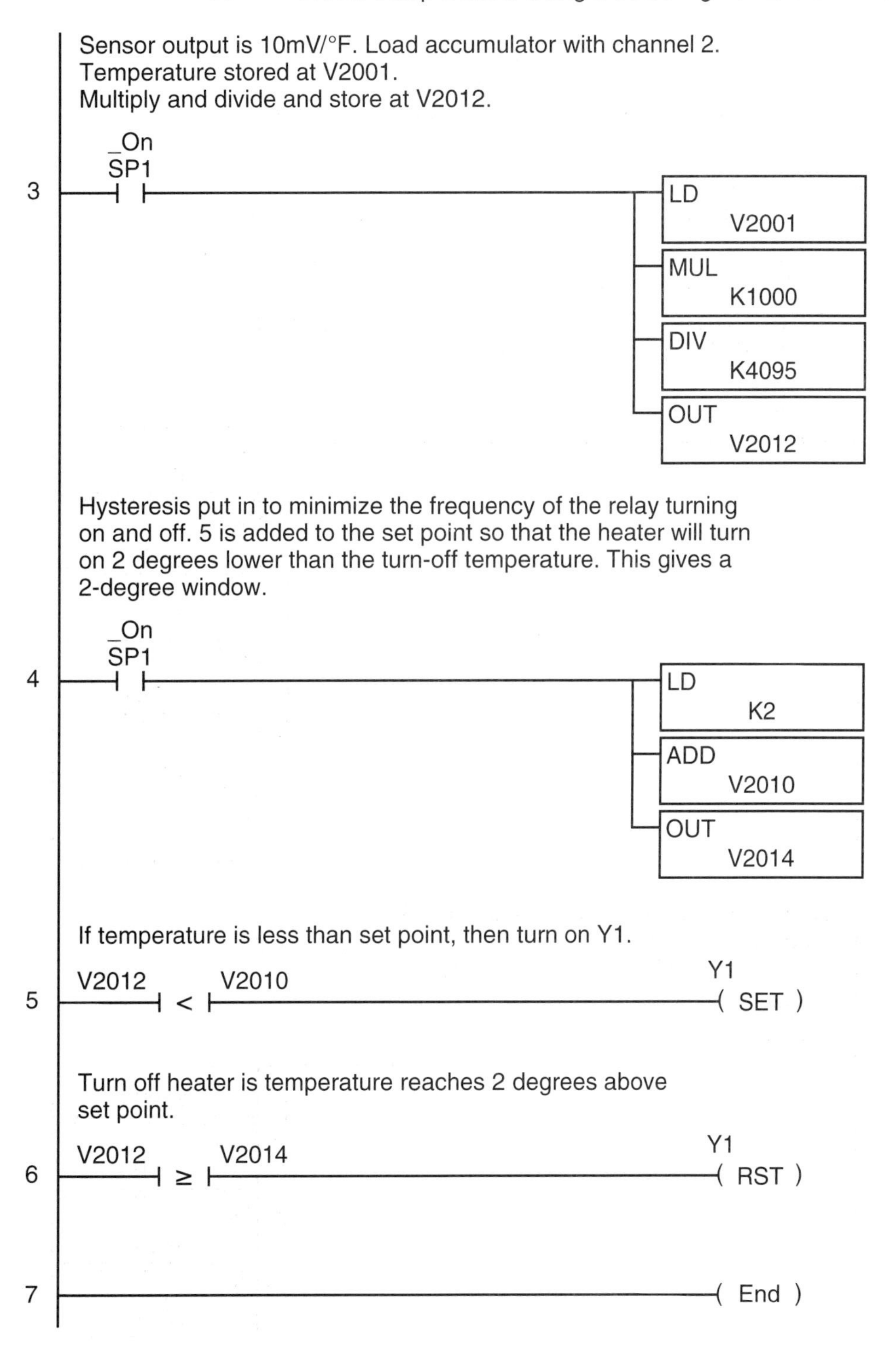

05 On/Off Temp Control Using a Soldering Iron and an ADC
Sensor output is 10mV/°F. Load accumulator with channel 2.
Temperature stored at V2001.
Multiply and divide and store at V2012.
_On
SP1
3
LD
V2001
MUL
K1000
DIV
K4095
OUT
V2012
Hysteresis put in to minimize the frequency of the relay turning on and off. 5 is added to the set point so that the heater will turn on 2 degrees lower than the turn-off temperature. This gives a 2-degree window.
_On
SP1
4
LD
K2
ADD
V2010
OUT
V2014
If temperature is less than set point, then turn on Y1.
5
V2012
<
V2010
Y1
SET
Turn off heater is temperature reaches 2 degrees above set point.
6
V2012
≥
V2014
Y1
RST
7
End

Figure 8.14b

Name ______________________________ Date ______________

Lab

Temperature Control with Time Proportioning

This lab demonstrates temperature control of a soldering iron using time proportioning. The temperature of the iron is controlled by varying the duty cycle of the iron by turning the iron on and off with a duty cycle that varies with the difference between the set point SP and the process variable PV (the measured temperature) according to the equation:

Duty Cycle = (SP–PV)/SP

The time the iron is on for is given by the equation:

ON TIME = ((SP – PV)/SP)*5s

The period of this cycle is set to 5 seconds.

The temperature is measured with an LM34 temperature sensor, which puts out 10mV/°F. It is connected to channel 2 of the ADC and a 10K pot to channel 1 of the ADC, which acts as the set point temperature. The soldering iron's tip is placed into a metal block along with the temperature sensor. A flow chart diagram illustrating the PLC program showing the function of each rung is shown in Figure 8.15. The wiring diagram and the program are illustrated in Figures 8.16 and 8.17.

To recap:

Input Channel 1—Input set point

Input Channel 2—Input temperature sensor

Y1—Output to soldering iron

Equipment Required

1. PLC with an ADC
2. 12V, 5V, and 24V power supply
3. LM34 temperature sensor
4. Soldering iron
5. Modified extension cord
6. 10K pot
7. A piece of metal to act as a thermal load

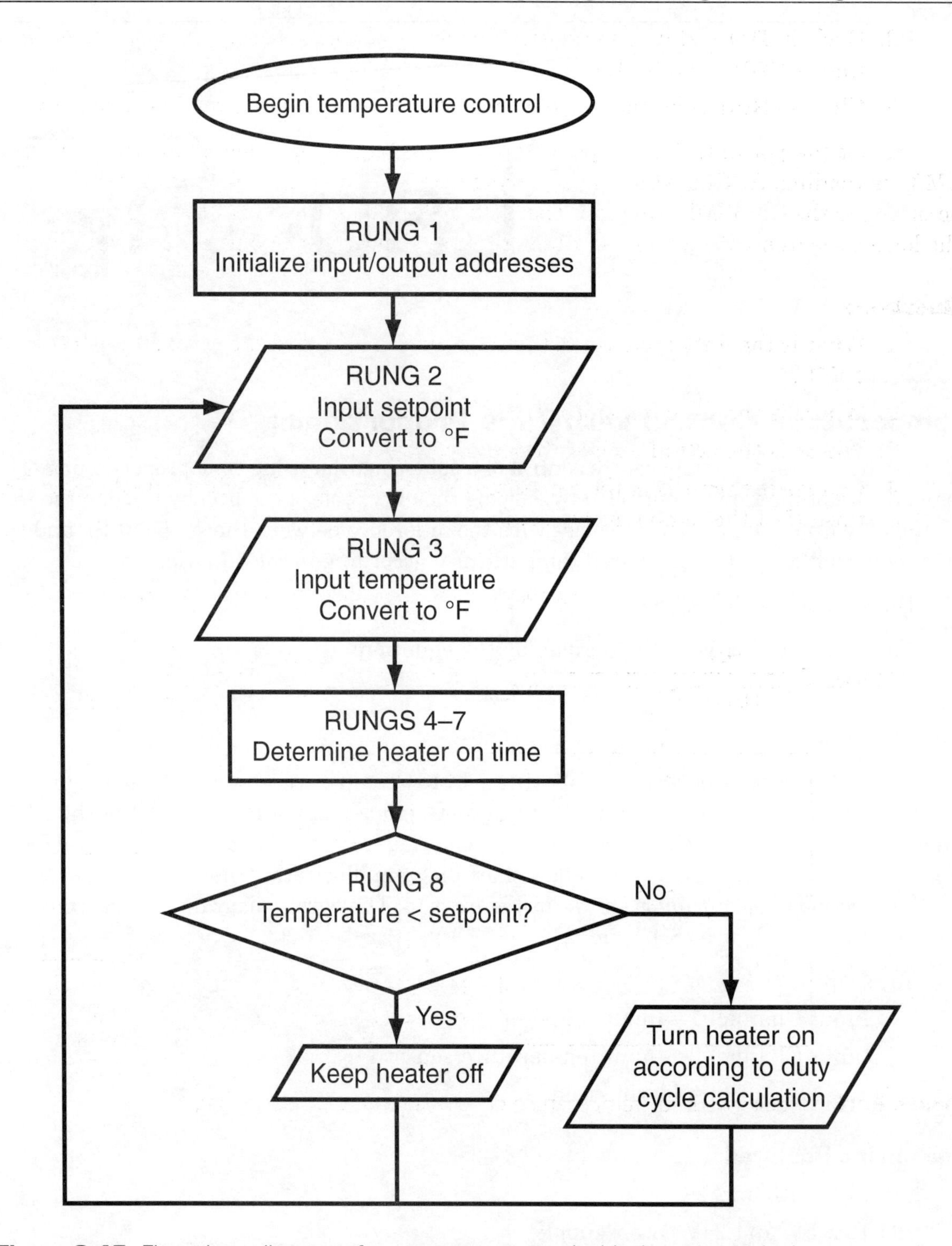

Figure 8.15 Flow chart diagram of temperature control with time proportioning.

To demonstrate the lab, connect up the circuit in Figure 8.16 and type in the program in Figure 8.17.

View the set point from channel 1 stored at V2010, and the temperature from channel 2 stored at V2011 by following these steps:

1. Click on **Debug** on the tool bar.

2. Click on **Data View** and type in the names of the variables you would like to see, that is V2010 and V2011.
3. Click on **Status** on the tool bar to see the current value of the variables listed.

Adjust the pot until the set point temperature is a little below the temperature the LM34 is reading. At this point the soldering iron should turn on for some period of time according to the ON TIME equation. The farther the temperature is away from the set point the longer the iron will stay on. At a temperature above the set point the iron will turn off.

Questions

1. What is the duty cycle when the set point is 150°F and the actual temperature is 100°F?
2. Where is the set point stored?
3. Where is the actual temperature stored?
4. Change the time 5s in the equation ON TIME = ((SP – PV)/SP)*5s to 3s. How does this affect the temperature regulation?

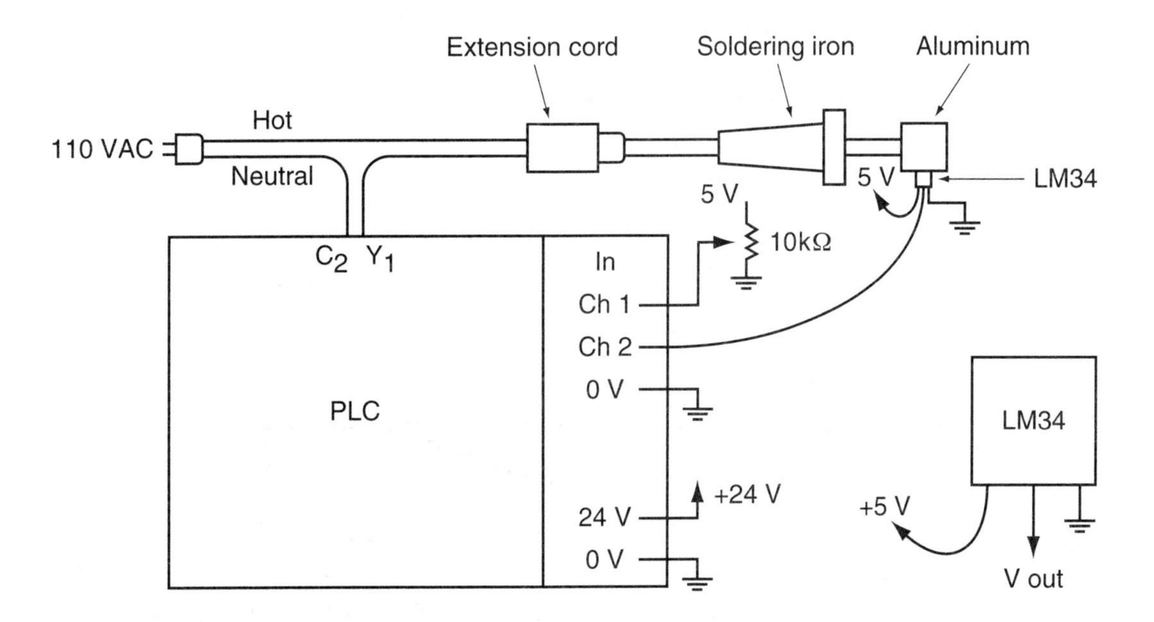

Figure 8.16 Circuit layout for temperature control with time proportioning.

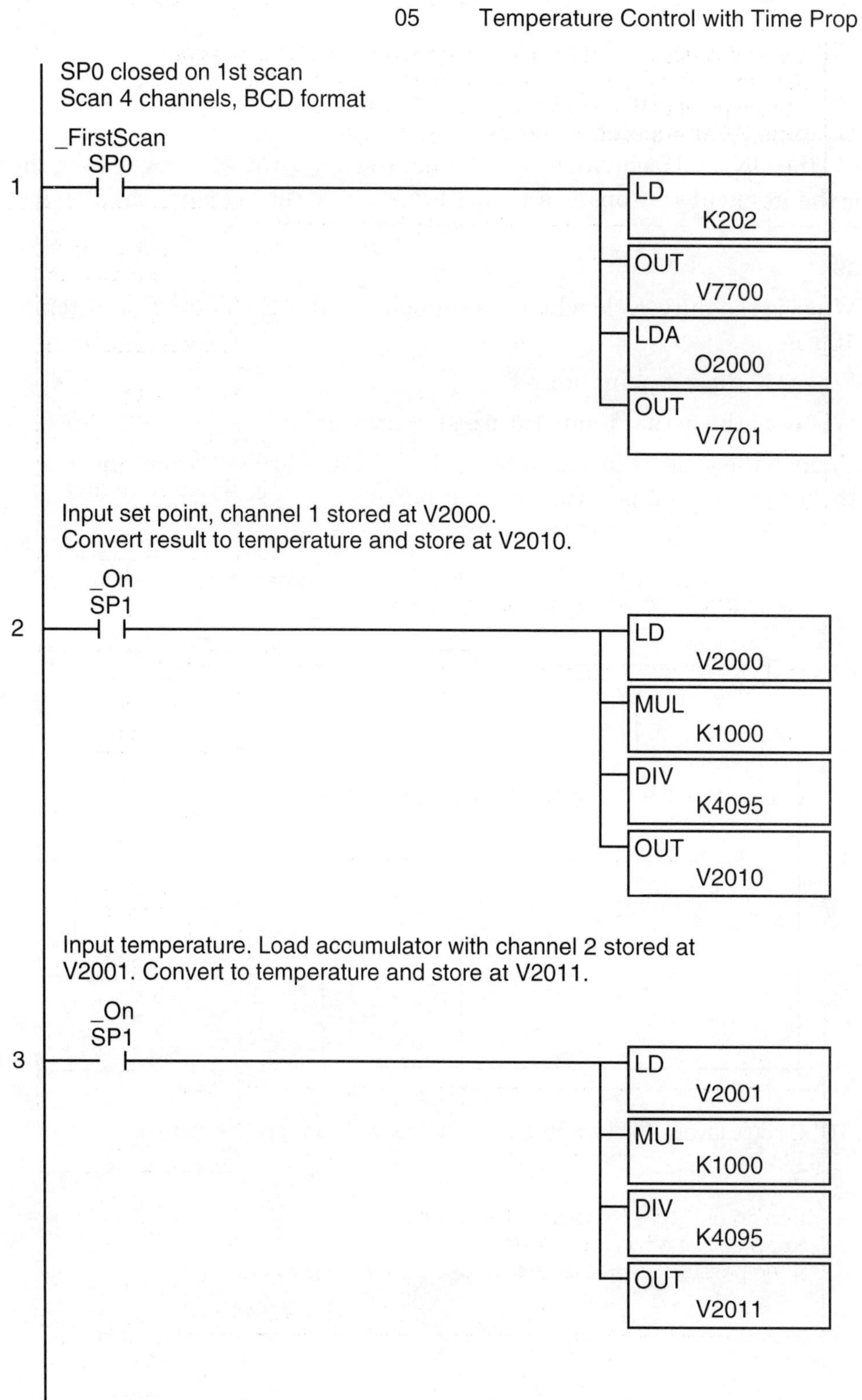

05 Temperature Control with Time Prop

SP0 closed on 1st scan
Scan 4 channels, BCD format

_FirstScan
SP0

1

LD K202
OUT V7700
LDA O2000
OUT V7701

Input set point, channel 1 stored at V2000.
Convert result to temperature and store at V2010.

_On
SP1

2

LD V2000
MUL K1000
DIV K4095
OUT V2010

Input temperature. Load accumulator with channel 2 stored at V2001. Convert to temperature and store at V2011.

_On
SP1

3

LD V2001
MUL K1000
DIV K4095
OUT V2011

Figure 8.17a

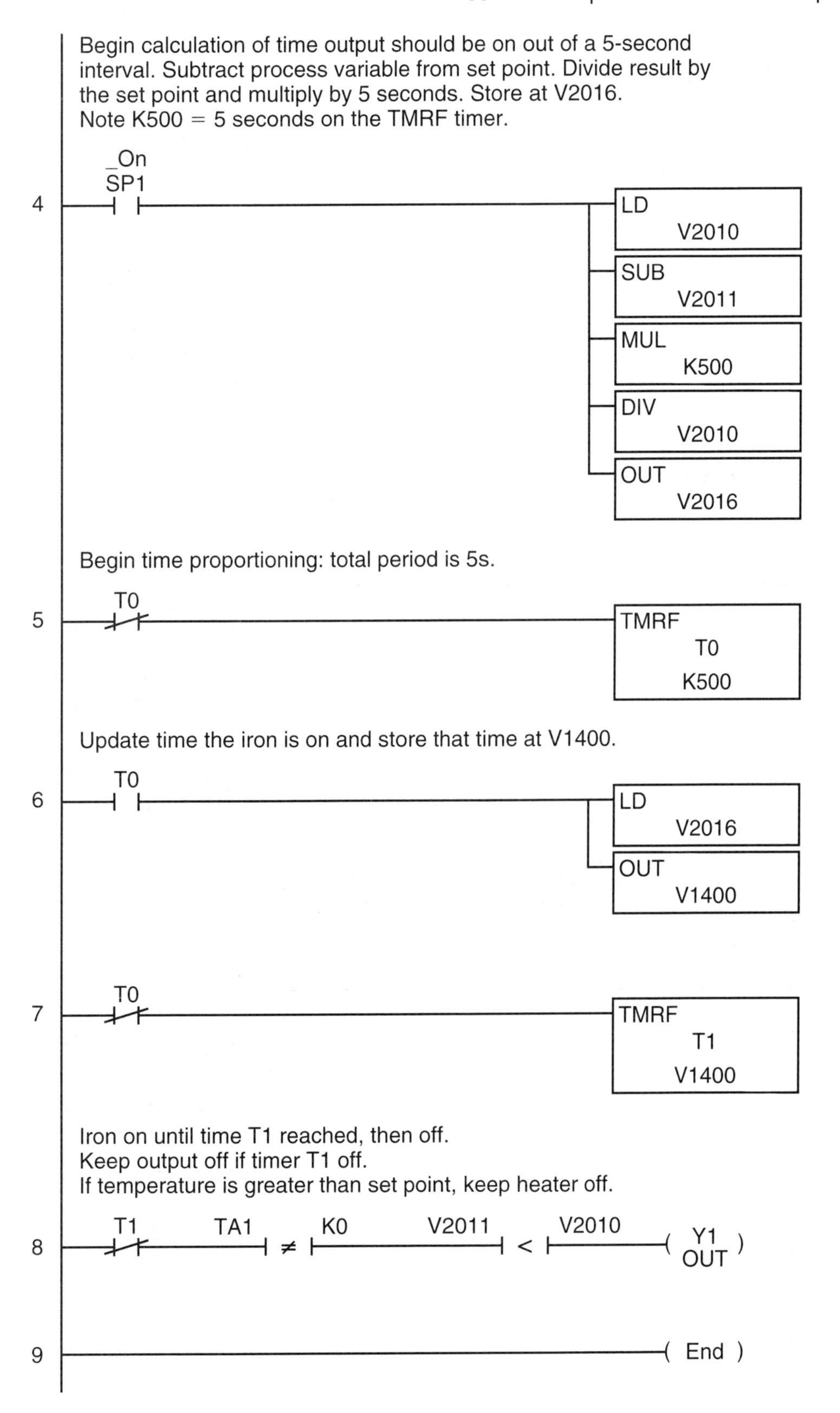
05 Temp Control with Time Prop
Begin calculation of time output should be on out of a 5-second interval. Subtract process variable from set point. Divide result by the set point and multiply by 5 seconds. Store at V2016.
Note K500 = 5 seconds on the TMRF timer.
4
_On
SP1
LD V2010
SUB V2011
MUL K500
DIV V2010
OUT V2016
Begin time proportioning: total period is 5s.
5
T0
TMRF T0 K500
Update time the iron is on and store that time at V1400.
6
T0
LD V2016
OUT V1400
7
T0
TMRF T1 V1400
Iron on until time T1 reached, then off.
Keep output off if timer T1 off.
If temperature is greater than set point, keep heater off.
8
T1
TA1
≠
K0
V2011
<
V2010
Y1 OUT
9
End

Figure 8.17b

Name ______________________________ Date ______________

Lab 53

Write Temperature Control Using Time Proportioning

Write a program to perform temperature control using time proportioning. Time proportioning is used to control temperature by turning on a fan for varying times depending on how far the actual temperature is away from the set point. As a source of heat, use a soldering iron placed nearby but not touching a mass that the sensor is placed into. Using the equation:

ONTIME=((SP – PV)/SP)*5s

where:

SP is the set point

PV is the actual temperature

Vary the number 5 in the equation and in the program in Lab 52. Change MUL K500 and observe how the temperature controller responds. Refer to Figures in Lab 50 and Lab 52 for help.

To recap:

Input Channel 1—Input set point

Input Channel 2—Input temperature sensor

Y1—Output to fan

Equipment Required

1. PLC with an ADC
2. 12V, 5V, and 24V power supply
3. LM34 temperature sensor
4. Soldering iron
5. 10K pot
6. 12V motor with some type of fan blade.

Name ______________________ Date ______________

Lab 

Proportional Temperature Control with an ADC and a DAC

This lab demonstrates proportional temperature control by controlling the speed of a fan to regulate temperature. The temperature is measured using an LM34 temperature sensor, which puts out 10mV/°F. It is connected to channel 2 of the ADC and a 10K pot connected to channel 1 of the ADC, which acts as the set point temperature. See Figure 8.19. The output to the fan will vary proportionally to the error between the set point SP and the actual temperature or "process variable" PV. The output from the DAC is connected to channel 1, which drives a power transistor which powers a fan. The proportional control is described by the following equation.

Output = Kp* Error

where:

Kp = Proportional gain

Error = SP–PV

SP = Set Point

PV = Process Variable

We will start with a gain of 200 and then vary it. This will result in a 5% change in the output for a 1° error because 200/4095 = 5%. A flow chart diagram illustrating the PLC program showing the function of each rung is shown in Figure 8.18. The wiring diagram and the program are illustrated in Figures 8.19 and 8.20.

To recap:

Input Channel 1—Input set point

Input Channel 2—Input temperature sensor

Output Channel 1—Fan circuit

Equipment Required

1. PLC with an ADC/DAC
2. 5V, 12V, and 24V power supply
3. LM34 temperature sensor
4. 10K pot

5. 2.2K resistor
6. 12V motor
7. 2N3054 power transistor

To demonstrate the lab, connect up the circuit in Figure 8.19 and type in the program in Figure 8.20.

View the set point from channel 1 stored at V2012, and the temperature from channel 2 stored at V2013 by performing the following steps:

1. Click on **Debug** on the tool bar.

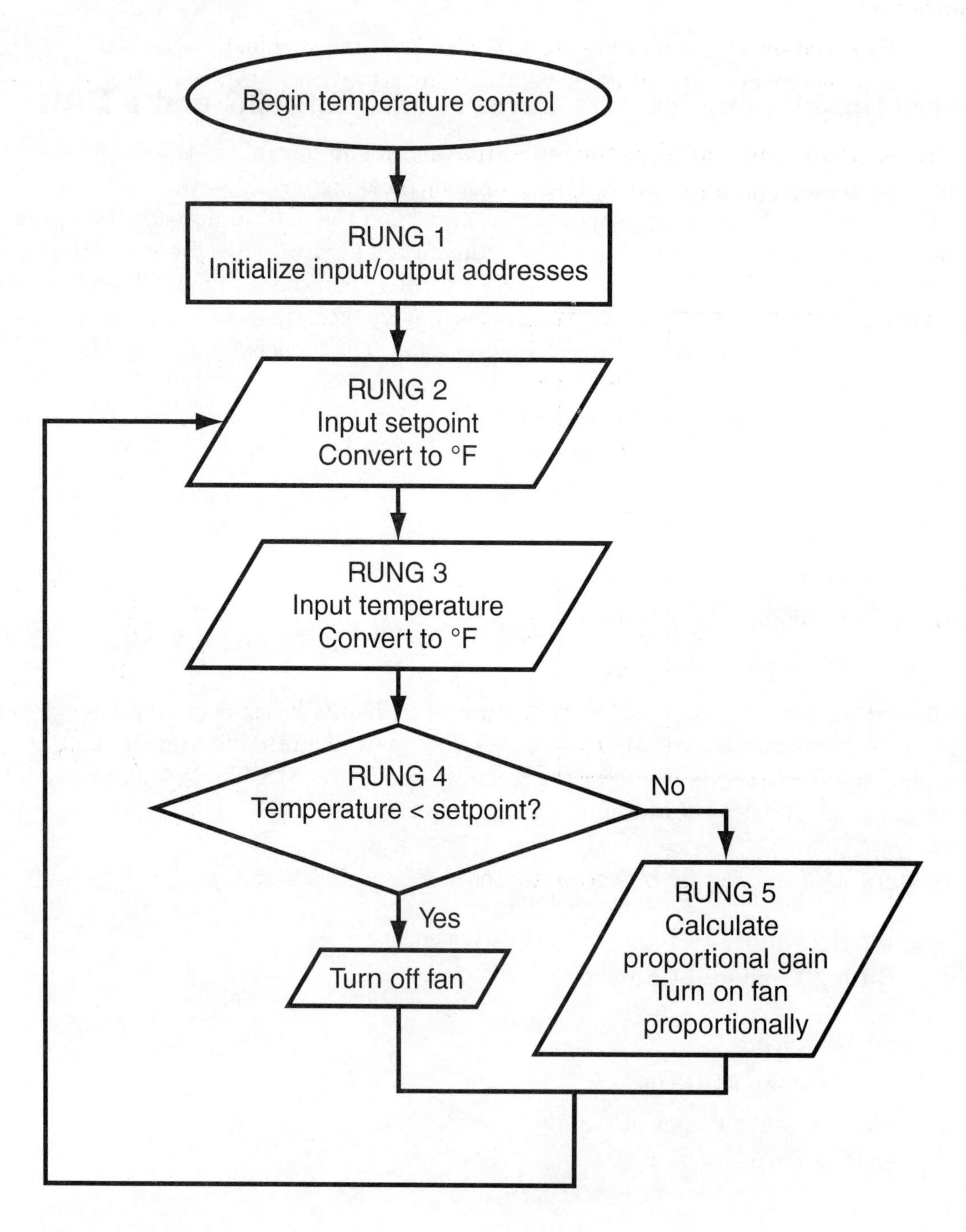

Figure 8.18 Flow chart diagram of proportional temperature control with an ADC and a DAC.

2. Click on **Data View** and type in the names of the variables you would like to see that is, V2012 and V2013.
3. Click on **Status** on the tool bar to see the current value of the variables listed.

Adjust the pot until the set point temperature is a little above the temperature the LM34 is reading. Place your fingers on the LM34 to raise its temperature above the set point. At this point, the fan should turn on. Note how fast the fan is spinning. The farther the temperature is above the set point, the faster the fan will spin. This is proportional control at work!

Questions

1. View the set point and temperature with Data View. Adjust the set point to slightly above room temperature, say 75°F. If the actual temperature is 80°F, what is the output?
2. At what temperature is the fan at maximum speed?
3. How much does the temperature vary about the set point?

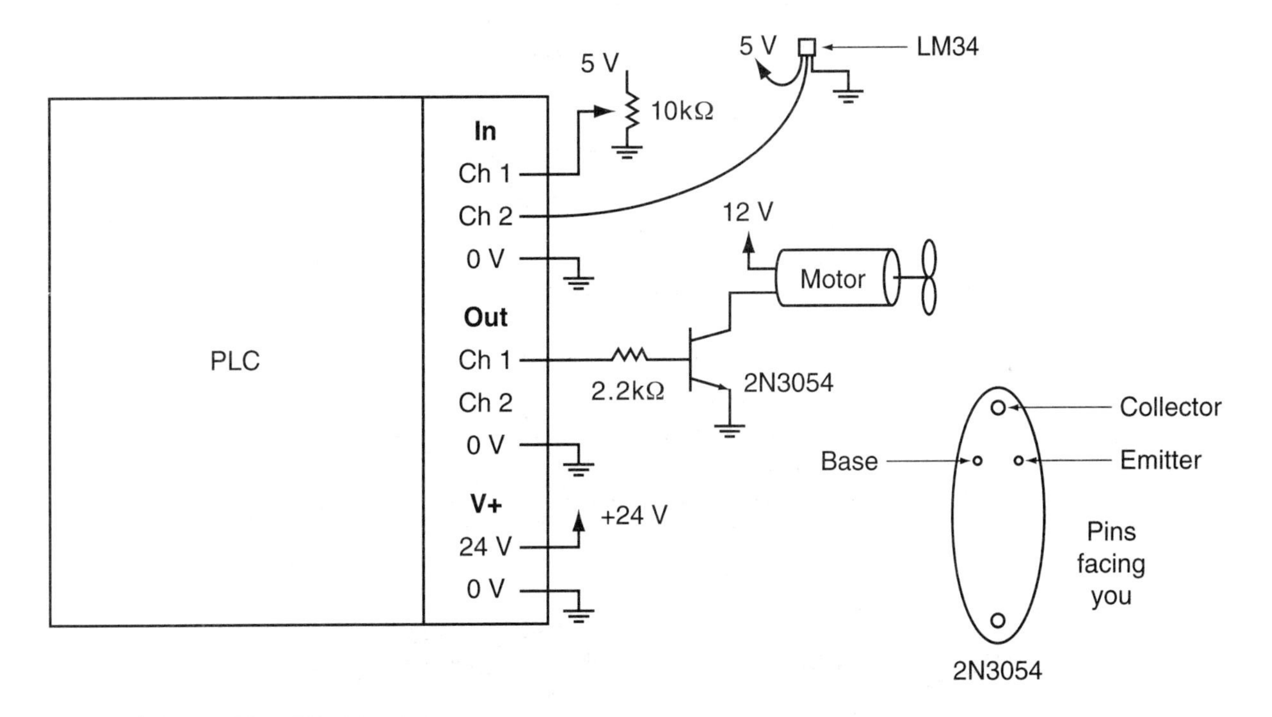

Figure 8.19 Circuit layout for proportional control with an ADC and a DAC.

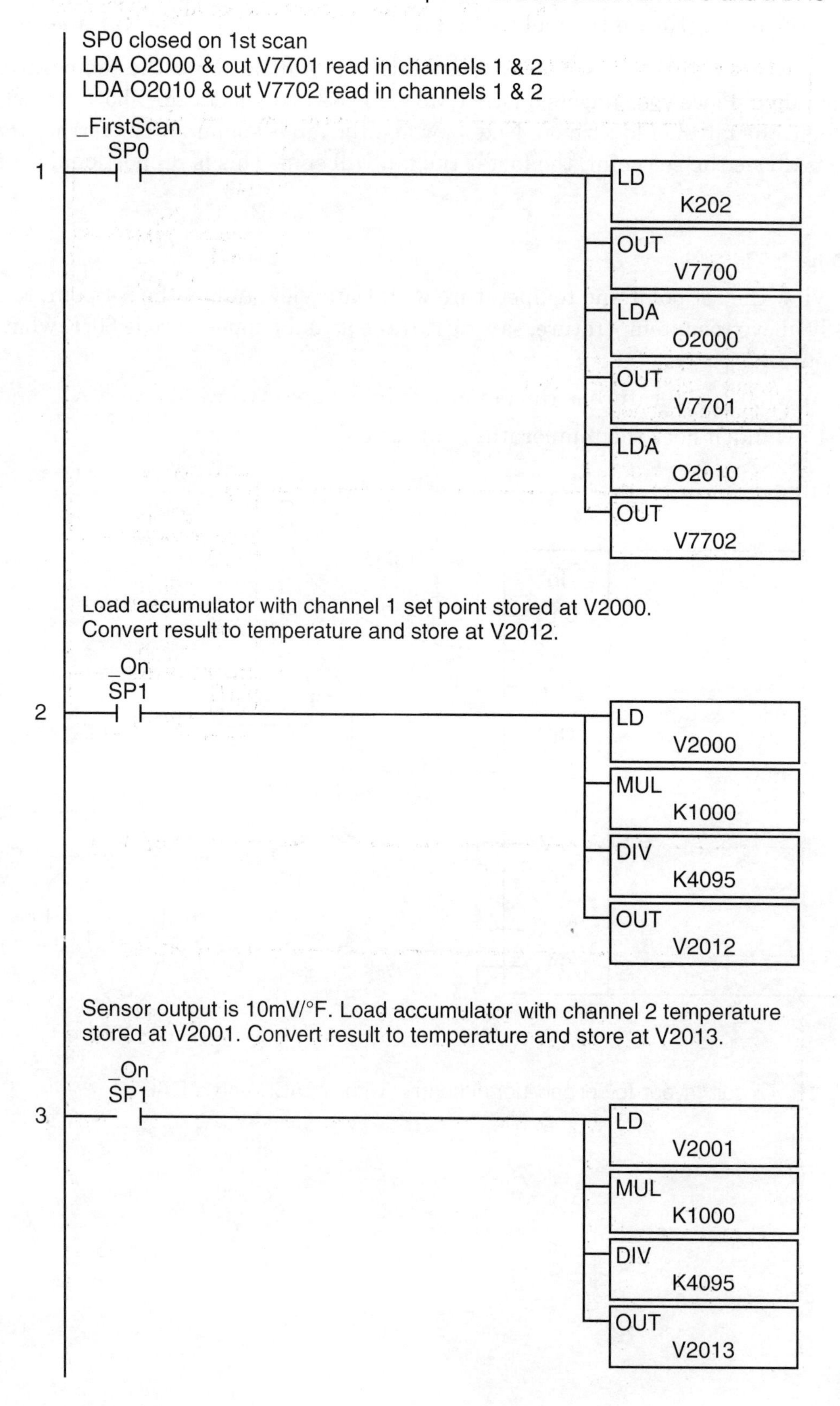
05 Proportional Control with an ADC and a DAC
SP0 closed on 1st scan
LDA O2000 & out V7701 read in channels 1 & 2
LDA O2010 & out V7702 read in channels 1 & 2
_FirstScan
SP0
1
LD
K202
OUT
V7700
LDA
O2000
OUT
V7701
LDA
O2010
OUT
V7702
Load accumulator with channel 1 set point stored at V2000.
Convert result to temperature and store at V2012.
_On
SP1
2
LD
V2000
MUL
K1000
DIV
K4095
OUT
V2012
Sensor output is 10mV/°F. Load accumulator with channel 2 temperature
stored at V2001. Convert result to temperature and store at V2013.
_On
SP1
3
LD
V2001
MUL
K1000
DIV
K4095
OUT
V2013

Figure 8.20a

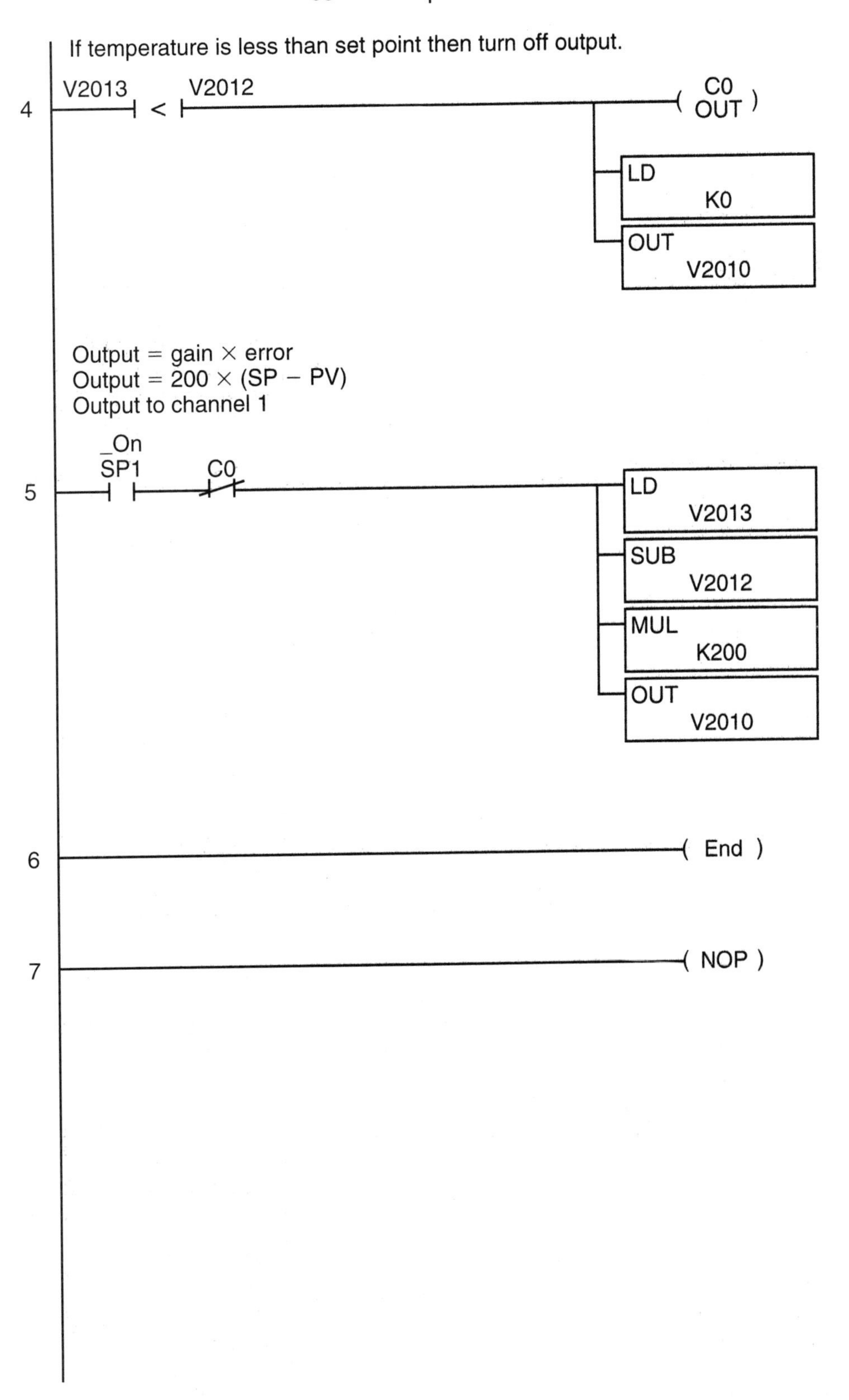
05 Proportional Control with an ADC and a DAC
If temperature is less than set point then turn off output.
4
V2013
<
V2012
C0
OUT
LD
K0
OUT
V2010
Output = gain × error
Output = 200 × (SP − PV)
Output to channel 1
_On
SP1
C0
5
LD
V2013
SUB
V2012
MUL
K200
OUT
V2010
6
End
7
NOP

Figure 8.20b

References

For PLC hardware and software:

1. AutomationDirect
   http://www.automationdirect.com/
2. Allen Bradley
   http://www.ab.com/plclogic/micrologix/
3. GE Fanuc VersaMax Micro
   http://www.geindustrial.com/cwc/gefanuc/support/plcio.htm
4. Siemens
   http://www.sea.siemens.com/controls/product/s7200/CNs7200.htm

APPENDIX B

Spec Sheets

DL05 I/O Specifications

D0-05DR $99.00

Wiring diagram and specifications

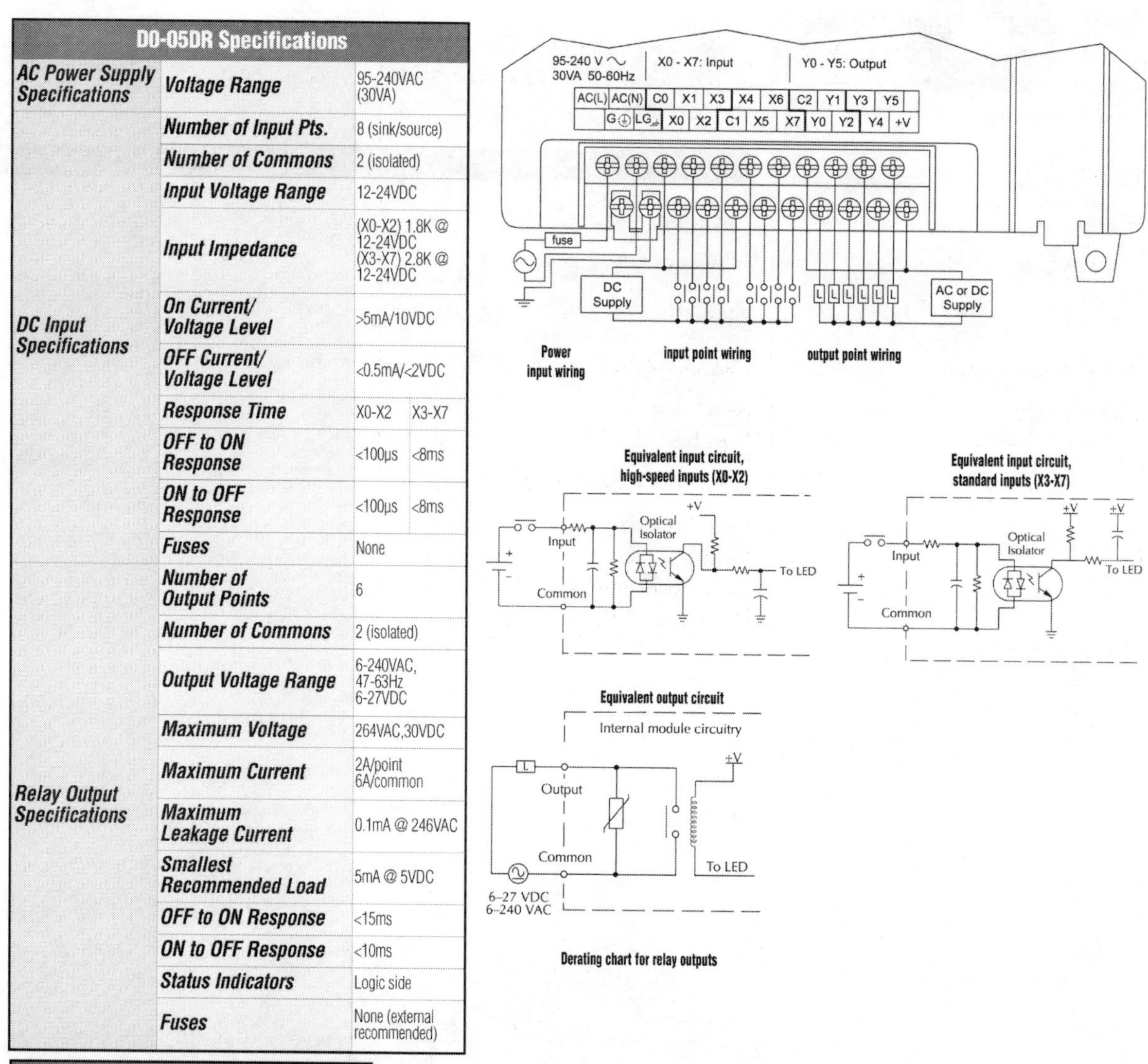

D0-05DR Specifications		
AC Power Supply Specifications	Voltage Range	95-240VAC (30VA)
DC Input Specifications	Number of Input Pts.	8 (sink/source)
	Number of Commons	2 (isolated)
	Input Voltage Range	12-24VDC
	Input Impedance	(X0-X2) 1.8K @ 12-24VDC (X3-X7) 2.8K @ 12-24VDC
	On Current/ Voltage Level	>5mA/10VDC
	OFF Current/ Voltage Level	<0.5mA/<2VDC
	Response Time	X0-X2 X3-X7
	OFF to ON Response	<100μs <8ms
	ON to OFF Response	<100μs <8ms
	Fuses	None
Relay Output Specifications	Number of Output Points	6
	Number of Commons	2 (isolated)
	Output Voltage Range	6-240VAC, 47-63Hz 6-27VDC
	Maximum Voltage	264VAC,30VDC
	Maximum Current	2A/point 6A/common
	Maximum Leakage Current	0.1mA @ 246VAC
	Smallest Recommended Load	5mA @ 5VDC
	OFF to ON Response	<15ms
	ON to OFF Response	<10ms
	Status Indicators	Logic side
	Fuses	None (external recommended)

Typical Relay Life (Operations) at Room Temperature		
Voltage and Type of Load	Load Current 1A	2A
24 VDC Resistive	600K	270K
24 VDC Solenoid	150K	60K
110 VAC Resistive	900K	350K
110 VAC Solenoid	350K	150K
220 VAC Resistive	600K	250K
220 VAC Solenoid	200K	100K

Figure B.1 Specifications on AutomationDirect's® DL05 PLC. Courtesy of AutomationDirect.com.

DL05/06 Option Modules

F0-2AD2DA-2 $149.00

2-point analog input and 2-point analog output module

F0-2AD2DA-2 Input Specifications	
Number of Channels	2, single ended (one common)
Input Range	0 to 5VDC or 0 to 10VDC (jumper selectable)
Resolution	12 bit (1 in 4096)
Step Response	10.0mS to 95% of full step change
Crosstalk	1/2 count max (-80db)*
Active Low-pass Filtering	-3dB at 300Hz (-12dB per octave)
Input Impedance	>20KΩ
Absolute Max Ratings	±15V
Linearity Error (end to end)	±2 counts (0.025% of full scale) max*
Input Stability	±1 count*
Gain Error	±6 counts max*
Offset Error	±2 counts max*
Max Inaccuracy	±0.3% at 25°C (77°F) ±0.6% at 0 to 60°-C (32 to 140°F)
Accuracy vs. Temperature	±100 ppm/°C typical

** One count in the specification table is equal to one least significant bit of the analog data value (1 in 4096)*

F0-2AD2DA-2 Output Specifications	
Number of Channels	2, single ended (one common)
Output Range	0 to5VDC or 0 to 10VDC (jumper selectable)
Resolution	12 bit (1 in 4096)
Conversion Settling Time	50μS for full scale change
Crosstalk	1/2 count max (-80db)*
Peak Output Voltage	±15VDC (power supply limited)
Offset Error	0.1% of range
Gain Error	0.4% of range
Linearity Error (end to end)	±1 counts (0.075% of full scale) max*
Output Stability	±2 counts*
Load Impedance	2KΩ max
Load Capacitance	0.01μF max
Accuracy vs. Temperature	±50 ppm/°C typical

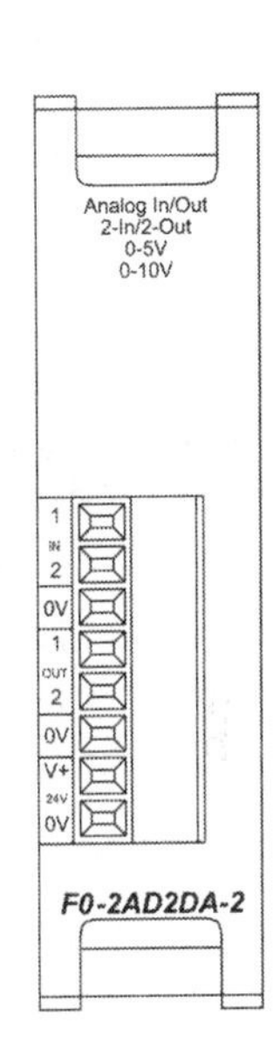

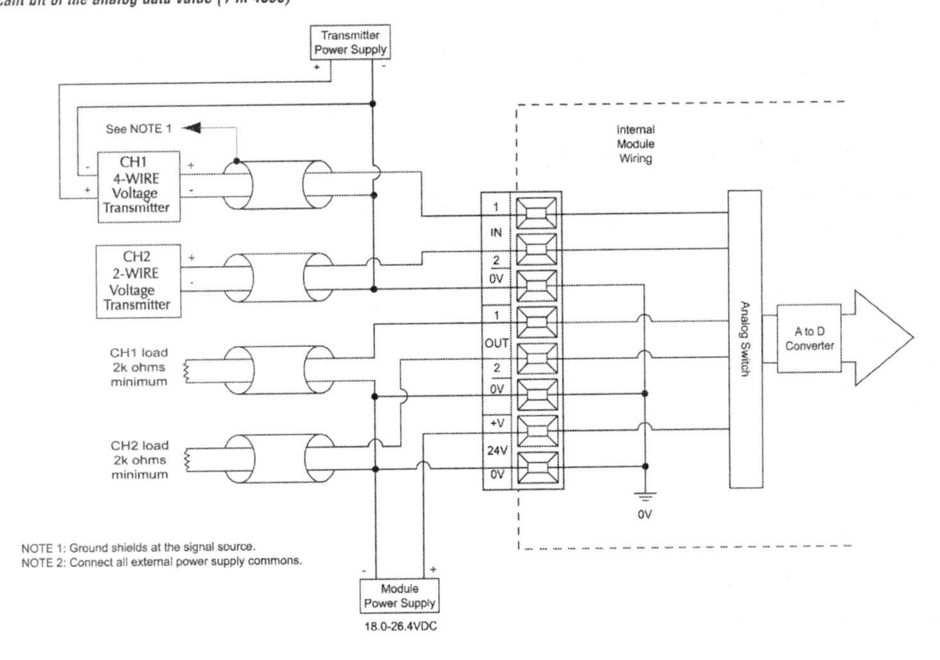

Figure B.2 Diagram of AutomationDirect's® Input/Output FO-2AD2DA-2 module. Courtesy of AutomationDirect.com.

August 2000

LM741
Operational Amplifier

General Description

The LM741 series are general purpose operational amplifiers which feature improved performance over industry standards like the LM709. They are direct, plug-in replacements for the 709C, LM201, MC1439 and 748 in most applications.

The amplifiers offer many features which make their application nearly foolproof: overload protection on the input and output, no latch-up when the common mode range is exceeded, as well as freedom from oscillations.

The LM741C is identical to the LM741/LM741A except that the LM741C has their performance guaranteed over a 0°C to +70°C temperature range, instead of −55°C to +125°C.

Connection Diagrams

Metal Can Package

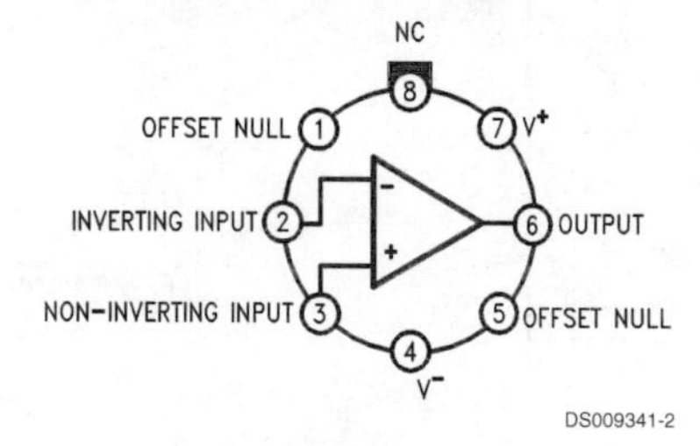

Note 1: LM741H is available per JM38510/10101

Order Number LM741H, LM741H/883 (Note 1), **LM741AH/883 or LM741CH**
See NS Package Number H08C

Dual-In-Line or S.O. Package

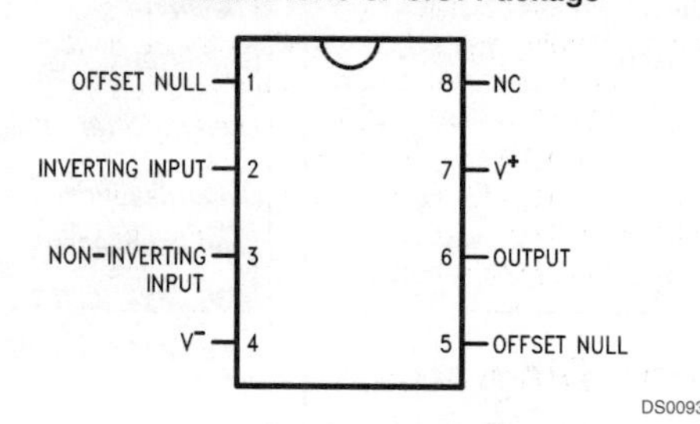

Order Number LM741J, LM741J/883, LM741CN
See NS Package Number J08A, M08A or N08E

Ceramic Flatpak

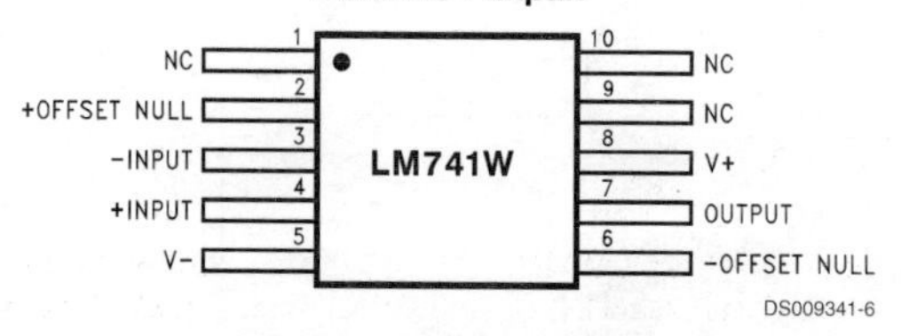

Order Number LM741W/883
See NS Package Number W10A

Typical Application

Offset Nulling Circuit

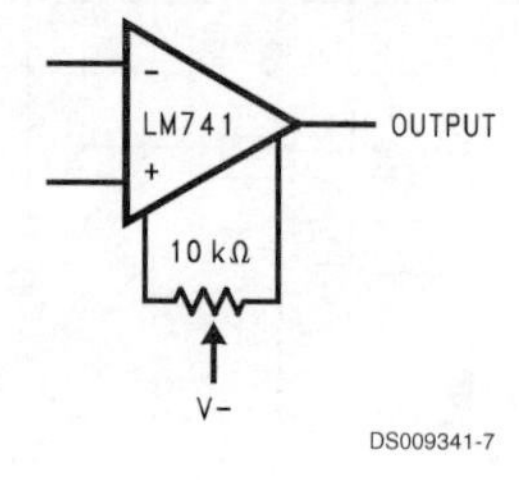

Figure B.3 The 741 op-amp. Courtesy of National Semiconductor.

November 2000

LM34
Precision Fahrenheit Temperature Sensors

LM34 Precision Fahrenheit Temperature Sensors

General Description

The LM34 series are precision integrated-circuit temperature sensors, whose output voltage is linearly proportional to the Fahrenheit temperature. The LM34 thus has an advantage over linear temperature sensors calibrated in degrees Kelvin, as the user is not required to subtract a large constant voltage from its output to obtain convenient Fahrenheit scaling. The LM34 does not require any external calibration or trimming to provide typical accuracies of ±½˚F at room temperature and ±1½˚F over a full −50 to +300˚F temperature range. Low cost is assured by trimming and calibration at the wafer level. The LM34's low output impedance, linear output, and precise inherent calibration make interfacing to readout or control circuitry especially easy. It can be used with single power supplies or with plus and minus supplies. As it draws only 75 µA from its supply, it has very low self-heating, less than 0.2˚F in still air. The LM34 is rated to operate over a −50˚ to +300˚F temperature range, while the LM34C is rated for a −40˚ to +230˚F range (0˚F with improved accuracy). The LM34 series is available packaged in hermetic TO-46 transistor packages, while the LM34C, LM34CA and LM34D are also available in the plastic TO-92 transistor package. The LM34D is also available in an 8-lead surface mount small outline package. The LM34 is a complement to the LM35 (Centigrade) temperature sensor.

Features

- Calibrated directly in degrees Fahrenheit
- Linear +10.0 mV/˚F scale factor
- 1.0˚F accuracy guaranteed (at +77˚F)
- Rated for full −50˚ to +300˚F range
- Suitable for remote applications
- Low cost due to wafer-level trimming
- Operates from 5 to 30 volts
- Less than 90 µA current drain
- Low self-heating, 0.18˚F in still air
- Nonlinearity only ±0.5˚F typical
- Low-impedance output, 0.4Ω for 1 mA load

Connection Diagrams

TO-46
Metal Can Package
(Note 1)

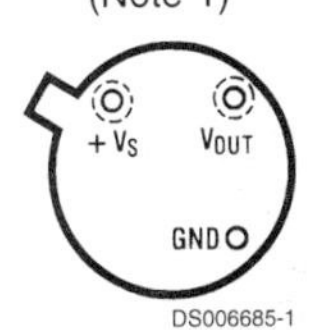

Order Numbers LM34H, LM34AH, LM34CH, LM34CAH or LM34DH
See NS Package Number H03H

TO-92
Plastic Package

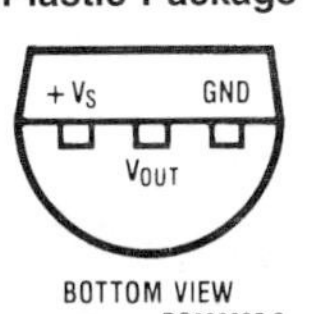

Order Number LM34CZ, LM34CAZ or LM34DZ
See NS Package Number Z03A

SO-8
Small Outline Molded Package

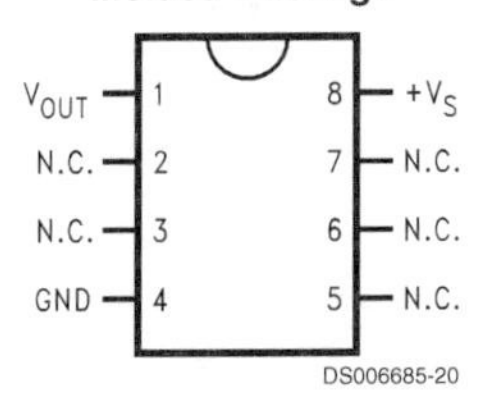

N.C. = No Connection

Top View
Order Number LM34DM
See NS Package Number M08A

Note 1: Case is connected to negative pin (GND).

 www.national.com

Figure B.4 The LM34 temperature sensor. Courtesy of National Semiconductor.

MINIATURE RELAY

DS2Y-RELAYS

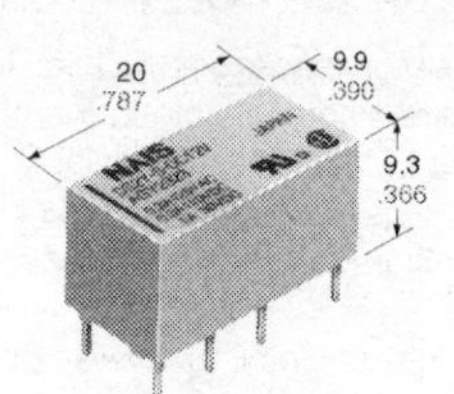

FEATURES

- **2 Form C contact**
- **High sensitivity-200 mW nominal operating power**
- **High breakdown voltage**
  1500 V FCC surge between open contacts
- **DIP-2C type matching 16 pin IC socket**
- **Sealed construction**

SPECIFICATIONS

Contact

Arrangement			2 Form C
Initial contact resistance, max. (By voltage drop 6 V DC 1 A)			50 mΩ
Contact material			Gold-clad sliver
Rating (resistive)	Max. switching power		60 W, 62.5 VA
	Max. switching voltage		220 V DC, 250 V AC
	Max. switching current		2 A
	Max. carrying current		3 A
Expected life (min. operations)	Mechanical		1×10^8
	Electrical	1 A 30 V DC	5×10^5
		2 A 30 V DC	1×10^5

Coil (polarized) (at 20°C 68°F)

Single side stable	Minimum operating power	Approx. 98 mW (147 mW: 48 V)
	Nominal operating power	Approx. 200 mW (300 mW: 48 V)
2 coil latching	Minimum set and reset power	Approx. 88 mW (177 mW: 48 V)
	Nominal set and reset power	Approx. 180 mW (360 mW: 48 V)

Remarks

- * Specifications will vary with foreign standards certification ratings.
- *1 Measurement at same location as "Initial breakdown voltage" section
- *2 Detection current: 10mA
- *3 Excluding contact bounce time
- *4 Half-wave pulse of sine wave: 11ms, detection time: 10μs
- *5 Half-wave pulse of sine wave: 6ms
- *6 Detection time: 10μs
- *7 Refer to 5. Conditions for operation, transport and storage mentioned in AMBIENT ENVIRONMENT (Page 61).

Characteristics (at 20°C 68°F)

Initial insulation resistance*1		Min. 100 MΩ (at 500 V DC)
Initial breakdown voltage*2	Between open contacts	750 Vrms
	Between contact sets	1,000 Vrms
	Between contact and coil	1,000 Vrms
FCC surge voltage between contacts and coil		1,500 V
Operate time*3 (at nominal voltage)		Approx. 4 ms
Release time*3 (at nominal voltage)		Approx. 3 ms
Set time*3 (latching) (at nominal voltage)		Approx. 3 ms
Reset time*3 (latching) (at nominal voltage)		Approx. 3 ms
Temperature rise		Max. 65°C with nominal voltage across coil and at nominal switching capacity
Shock resistance	Functional*4	Min. 490 m/s² {50 G}
	Destructive*5	Min. 980 m/s² {100 G}
Vibration resistance	Functional*6	10 to 55 Hz at double amplitude of 3.3 mm
	Destructive	10 to 55 Hz at double amplitude of 5 mm
Conditions for operation, transport and storage*7 (Not freezing and condensing at low temperature)	Ambient temp.	–40°C to +70°C –40°F to +158°F
	Humidity	5 to 85% R.H.
Unit weight		Approx. 4 g .14 oz

FCC (Federal Communication Commission) requests following standard as Breakdown Voltage specification.

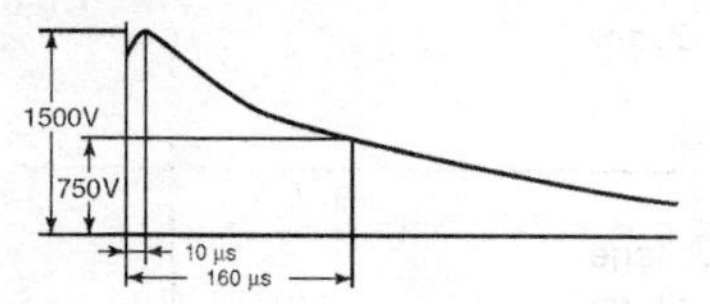

TYPICAL APPLICATIONS

- Telecommunication equipment
- Office equipment
- Computer peripherals
- Security alarm systems
- Medical equipment

ORDERING INFORMATION

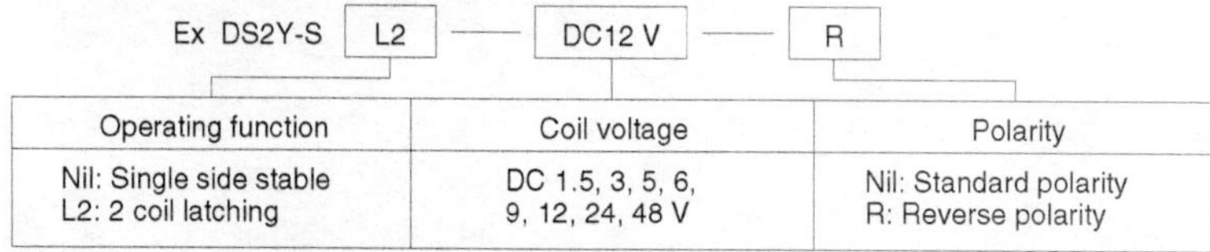

Ex DS2Y-S | L2 | — | DC12 V | — | R

Operating function	Coil voltage	Polarity
Nil: Single side stable L2: 2 coil latching	DC 1.5, 3, 5, 6, 9, 12, 24, 48 V	Nil: Standard polarity R: Reverse polarity

(Note) Standard packing: Carton: 50 pcs. Case: 500 pcs.

Figure B.5 A DPDT relay. Courtesy of Aromat.

H21A1 / H21A2 / H21A3
PHOTOTRANSISTOR OPTICAL INTERRUPTER SWITCH

PACKAGE DIMENSIONS

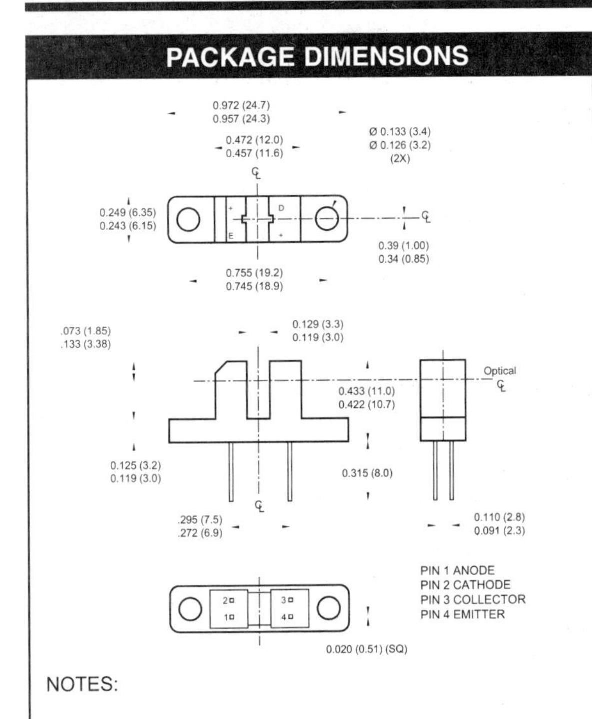

NOTES:

1. Dimensions for all drawings are in inches (mm).
2. Tolerance of ± .010 (.25) on all non-nominal dimensions unless otherwise specified.

DESCRIPTION

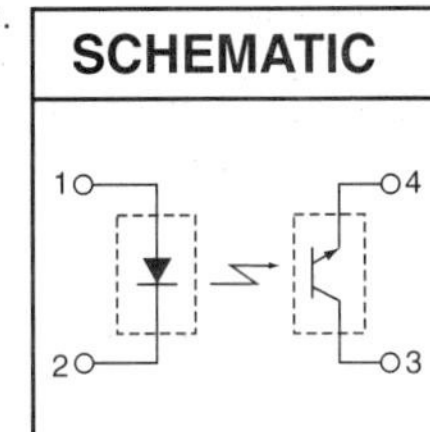

The H21A1, H21A2 and H21A3 consist of a gallium arsenide infrared emitting diode coupled with a silicon phototransistor in a plastic housing. The packaging system is designed to optimize the mechanical resolution, coupling efficiency, ambient light rejection, cost and reliability. The gap in the housing provides a means of interrupting the signal with an opaque material, switching the output from an "ON" to an "OFF" state.

SCHEMATIC

1 4
2 3

FEATURES

- Opaque housing
- Low cost
- .035" apertures
- High $I_{C(ON)}$

1. Derate power dissipation linearly 1.33 mW/°C above 25°C.
2. RMA flux is recommended.
3. Methanol or isopropyl alcohols are recommended as cleaning agents.
4. Soldering iron tip 1/16" (1.6mm) minimum from housing.

ABSOLUTE MAXIMUM RATINGS (T_A = 25°C unless otherwise specified)

Parameter	Symbol	Rating	Unit
Operating Temperature	T_{OPR}	-55 to +100	°C
Storage Temperature	T_{STG}	-55 to +100	°C
Soldering Temperature (Iron)(2,3 and 4)	T_{SOL-I}	240 for 5 sec	°C
Soldering Temperature (Flow)(2 and 3)	T_{SOL-F}	260 for 10 sec	°C
INPUT (EMITTER) Continuous Forward Current	I_F	50	mA
Reverse Voltage	V_R	6	V
Power Dissipation (1)	P_D	100	mW
OUTPUT (SENSOR) Collector to Emitter Voltage	V_{CEO}	30	V
Emitter to Collector Voltage	V_{ECO}	4.5	V
Collector Current	I_C	20	mA
Power Dissipation (T_C = 25°C)(1)	P_D	150	mW

Figure B.6 A slotted optical interrupter switch,. Courtesy of Fairchild Semiconductor.